THE

LONGEVITY

PLAN

*Seven Life-Transforming
Lessons from Ancient China*

DR. JOHN D. DAY

AND

JANE ANN DAY

with Matthew LaPlante

HARPER

NEW YORK · LONDON · TORONTO · SYDNEY

HARPER

This book contains advice and information relating to health care. It should be used to supplement rather than replace the advice of your doctor or another trained health professional. If you know or suspect you have a health problem, it is recommended that you seek your physician's advice before embarking on any medical program or treatment. All efforts have been made to assure the accuracy of the information contained in this book as of the date of publication. This publisher and the author disclaim liability for any medical outcomes that may occur as a result of applying the methods suggested in this book.

A hardcover edition of this book was published in 2017 by HarperCollins Publishers.

All photographs courtesy of the author.

HarperCollins books may be purchased for educational, business, or sales promotional use. For information, please email the Special Markets Department at SPsales@harpercollins.com.

FIRST HARPER PAPERBACK EDITION PUBLISHED 2018.

Designed by Fritz Metsch

Library of Congress Cataloging-in-Publication Data has been applied for.

ISBN 978-0-06-231982-1 (pbk.)

20 21 22   LSC   10 9 8 7 6 5 4 3

TO OUR
FAMILIES
———

THE

LONGEVITY

PLAN

———

CONTENTS

THE

LONGEVITY

PLAN

———

Prologue

LONGEVITY

SOUP

"COME NOW," SHE SAID. "I WILL TEACH YOU TO MAKE LONGEV-ity soup."

I'd been waiting for this moment for a very long time; finally I would learn the secret. Finally I'd know.

Feng Chun crushed a handful of hemp seeds, and then another. She strained them with hot water. She boiled the mixture in a wok, added some pumpkin greens, and stirred.

"There you are," she said, pouring the gray liquid into a bowl.

"Wait," I said. "That's it?"

"That's it."

"But . . . but that's so simple!"

"Of course it is," she said. "What did you expect?"

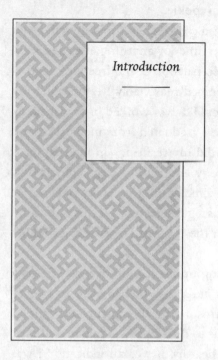

Introduction

BY MID-MORNING THE DOUGHNUTS WOULD BE GONE.

That fact was an essential part of my planning each day as I prepared my breakfast in the doctor's lounge at the hospital where I work. I'd always grab a doughnut, a bagel, and a Diet Coke. Then I'd grab a second doughnut, wrap it in a napkin, and stash it in a cabinet just outside of the operating room.

My colleagues laughed and rolled their eyes. I just shrugged. It all seemed perfectly rational to me.

My days as a cardiologist were filled with pacemaker implantations, procedures to three-dimensionally "map and zap" potentially fatal heart arrhythmias, and defibrillator surgeries. In between I'd snack.

Lunch on most days was a slice of pizza, or two, and another Diet Coke. On long days, I dined in the hospital cafeteria on a cheeseburger, fries, and a chocolate chip cookie.

I knew these weren't good food choices. But I told myself, given my hectic schedule, I didn't have time for anything else. Besides, I justified, many other doctors also partook of the free junk food at the hospital, and all of them seemed reasonably healthy. And my hospital was just like all the others I'd ever worked in or visited. At Johns Hopkins University where I graduated from medical school. At Stanford University where I did my residency in internal medicine and fellowships in cardiology and cardiac electrophysiology. As an assistant professor of medicine at the University of Utah. At nearly every hospital I'd visited as a guest lecturer. If this is the kind of food offered to doctors all over the country, I reasoned, it couldn't be *that* bad.

I always figured I was making up for it with exercise. I was a religious runner—a marathoner, no less—and had been for twenty years. It's one thing to eat healthy and be able to run 26.2 miles. I was eating *trash* and was still able to do it. Surely, I told myself, that wasn't just an indication I was healthy, it was an indication I was *more* than healthy.

Except I wasn't. Not even close.

It wasn't just what I ate; it was how I lived. I worked too many hours. I took too few vacation days. I didn't spend nearly enough time with my family. I spent a lot of time considering my productivity, and not much time contemplating my purpose. Life was a bit of a blur.

I was overweight, overworked, hypertensive, and had a cholesterol level much higher than it should have been. I was tired and stressed all the time.

I was also in constant pain. There was pain in my chest from acid reflux. There was pain in my back and neck from an autoimmune disease called ankylosing spondylitis. Food often became lodged in my esophagus from a condition called eosinophilic esophagitis, which made it difficult for me to swallow.

Many of these conditions ran in my family. And so I blamed my genes. I figured there wasn't much sense in trying to fight it. This was just part of getting old. This was my lot in life.

I took five medications daily. And that helped . . . a bit . . . for a while.

At forty-four years old, I found myself daydreaming about retirement. Someday I'd settle down and life would be good again. Or maybe it would just be less bad. That was the same thing, wasn't it?

In the meantime: One more busy week. One more missed vacation. One more doughnut.

I DON'T PARTICULARLY enjoy talking about the way I was back then, but my hope is that, in coming clean about my challenges, you'll come to see that the health struggles you've faced in your life can be resolved with a few minor course corrections. Forgive me if I shed a bit of modesty here to drive this point home, but I'm a good doctor. I've recently completed my term as president of the Heart Rhythm Society, an international organization of thousands of cardiologists in more than seventy countries. Over my twenty-plus-year career I've performed more than 6,000 catheter ablations and more than 3,000 pacemaker or defibrillator implantations. I've treated tens of thousands of patients.

I had access to more information about healthy living than most people could ever dream of, and all the resources I needed to make changes. In spite of all of that, I was still confused about what I should be doing to get myself back on track to a happier and healthier life. So whether this is the first time you've ever considered making changes to your life to improve your health or you've been trying for years, you're in good company.

And the truth is that even though I've turned my life around in a way that feels to me and my family like a miracle, I'm not here to peddle miracles, least of all by telling you that you should do everything I did, because it doesn't work that way. Everyone's a bit different, and some of us are a lot different. So what I'd like to do is help you figure out what works for *you*. Regardless of our individual circumstances, there do exist basic principles of well-being that can lead us all to a better life, but you get to choose how to adapt these principles in your own journey.

And on that journey, I'd be pleased to be your guide.

Not by myself, though. In these pages, I'm going to introduce you to some of the world's most qualified people on the subject of living longer, healthier, and happier lives. Their names are Boxin, Magan, Maxue, Mawen, Masongmou, Makun, and Makang. In 2012, they were the seven centenarians of Bapan, a village in southwest China, not far from the Vietnamese border, that rests in the middle of a region with one of the highest known concentrations of people over the age of one hundred anywhere in the world. These six women and one man, along with countless others, have lived by these basic principles of well-being without ever thinking about it. It's simply part of their lives.

I'm not only going to tell you how they live today, because no one wants to live like a centenarian, no matter how healthy they might be. I'm also going to tell you how they lived *throughout* their lives. I'll also introduce you to some of the other people, from every generation, who live, laugh, love, and work in this remarkable place. Together, these people have helped me shape my ideas on well-being, and those ideas, in turn, have helped me help lots of my patients *be well*.

In 2014, I began a series of four-month support groups comprising patients who worked together to apply the lessons of Longevity Village to their lives. Even having come to believe strongly in the power of the Longevity Village lifestyle, I was astonished by the results; 92 percent of the participants were able to adhere to their plans and stay on pace to reach their health goals. These are people who had abused their bodies for years, had decades upon decades of bad health habits, and often had no real support at home. Despite these challenges, most have been able to reverse at least some of their chronic medical conditions, including diabetes, hypertension, obesity, atrial fibrillation, insomnia, fatigue, acid reflux, heart failure, and high cholesterol.

I've seen similar outcomes among hundreds of other individual patients who have embraced these lessons. After launching a website dedicated to helping people live happier and healthier lives, people from around the world have shared with me their stories of radical personal transformation. And, of course, my own life stands in testament to the effectiveness of this model; it has been completely transformed.

Why does it work? Janine, a forty-one-year-old programmer from San Francisco, was battling obesity and some associated heart irregularities when she first came to see me. In nine months, her weight was down more than 45 pounds and her heart troubles were sub-diagnostic, as though they'd never occurred at all. "For me," she wrote six months into her Longevity Village journey, "this way of living just feels right. It's hard to explain, but it's almost like this is the way we would all be living if our ancestors had just recognized that, as we modernized, we couldn't simply leave everything that was good about the old ways behind."

With those words, Janine eloquently shared something I'd had a bit of trouble expressing when I was first explaining this health model to my patients. The Longevity Village lifestyle isn't about living like people in a remote part of China did in the past; it's about living in the modern world with a bit of ancient wisdom to guide our way toward happier, healthier futures.

I CAME TO learn about Bapan almost by accident. When I was nineteen years old, as part of my faith, I'd spent two years working with the Chinese immigrant population in New York City. Until that point, I didn't know the first thing about China. I didn't know anything about its rich history or cultural traditions. I didn't know a word of Mandarin. I didn't even like Chinese food.

But during that amazing time in my life, I came to adore the language, the culture, and the people I lived with and worked among. Long after I returned home to Utah from New York, I remained fascinated by China, and continued working to develop my language skills, such that today I am one of few Caucasian doctors who regularly gives medical lectures in Chinese. I'm told my accent isn't half bad. "You're like a proper *Běijīng rén*," a friend from China's capital city told me recently, using the words that describe a resident of Beijing. I beamed with pride.

Mandarin isn't an easy language to master, though, and thirty years after first learning how to say *nǐ hǎo* with appropriate intonation, I'm still working on it. So each week, over a video conference call, I meet with my Mandarin language coach, Zheng Lv, who lives in Xi'an, the starting point of the northern route of the famed Silk Road and the home of Emperor Qin Shi Huang's incredible Terracotta Army. As we chat, Zheng helps correct my tones and pronunciation, and sometimes introduces me to new words and Chinese concepts. Our sessions together are generally conversations, prompted by something we've heard about in the Chinese or American news media, and sometimes I tell her about an article of particular interest I've read in a Chinese or American medical journal.

That's what happened in 2012, when I mentioned to Zheng an article I'd read about the Bama County Centenarian Study, which had been published in a Chinese medical journal. At the time I was exploring the literature on how certain groups of people, living in certain ways, seem to be defying the conventional laws of aging. When I mentioned the article, Zheng told me she'd just seen a TV program about this region of China, where people reportedly live remarkably long lives free of the conditions that typically come with aging. The village of Bapan, Zheng said, was getting quite a bit of attention in China. "They say the land has magical properties," she told me. "In China they now call this place Longevity Village."

Longevity Village, I learned, was a small, poor, and remote town of just a few hundred people in the Guangxi Zhuang Autonomous Region. I'd never been to that part of China before, but I knew that rural villages in China's more remote areas often suffered from a lack of quality medical services. I also knew that, in general across the world, poor people don't tend to live as long and often have poorer health than people who live in more developed areas. Yet if what I was learning was true, none of that seemed to matter. Through happiness and hardship, into their eighties, nineties, and one hundreds, with few modern conveniences and even less medical intervention, the people of Bapan survive and thrive.

Over time, Zheng and I would return to the subject of Bapan again and again. I'd tell her what I'd been learning in my studies of the medical literature, and she would tell me what she was hearing about this area in China's popular media. I felt like I simply couldn't get enough information.

"*Qǐng duō gàosù wǒ yīdiǎn*," I'd say to her. "Please, tell me more."

Zheng probably wondered why I was so obsessed. What she didn't know was that I'd finally hit a health crisis and the solutions I'd tried simply weren't working. My conditions had worsened. My pain had, too. I couldn't run like I used to, so I was putting on even more weight. At the end of each day at the hospital I felt exhausted, but at night I was restless. I was tired all the time, so I was irritable.

I'd lost hope.

Bit by bit, though, I was finding glimpses of it in what I was learning about Longevity Village, and every time I'd find a new bit of research, or found another doctor who had done work in Bama County, I'd feel as though I was further unlocking some sort of magical treasure chest. For most of my life I've been an early-to-bed sort of guy, but I spent long hours, late into the night, hovered over my computer, poring over the Chinese medical literature in search of more information about Bapan.

It was my wife, who quickly came to share my excitement, who finally convinced me we needed to go.

So that's what I did.

And it changed my life.

MY FIRST TRIP to Bapan came in the summer of 2012. With me, as she has been on all our excursions to Bama County, was my wife Jane. Joining us was our eldest son, Joshua, who was then nine years old.

We'd intended to arrive in the village the evening before, rest up, and head straight to the home of one of the village centenarians at first daylight. Getting to Longevity Village had proved to be a challenge, though. We'd faced torrential rains on narrow mountain roads as we moved deep into northwest Bama County only to find, as night fell and the thunder and lightning pounded the skies, that we'd been dropped off at the wrong village. We stayed in a guesthouse and, when we awoke the next morning, learned we were not far from Bapan. We made our way there in a rickety three-wheeled moto-taxi, which dropped us off on the main road.

I probably should have been tired after such a long trip. It had been a three-day journey from our home in the United States and, troubled by the notion that we'd already been steered off course, I hadn't slept well the night before. But as we stepped into the village and saw a welcome sign festooned with the photos of the village's seven centenarians, I felt a surge of energy and excitement unlike anything I'd ever experienced. And as I looked around at my traveling companions it was clear they felt the same.

Underneath each photograph was a brief biography of each of the elders in Chinese characters. I translated the words for my wife and son.

"Some of these people were here a hundred years before I was even born!" Joshua marveled.

"Who should we meet first?" I asked.

"I'm dying to meet Boxin," Jane replied, pointing to the weather-faded photograph of the man at the center of the sign. "Can we find him first?"

Boxin, pronounced (bwo-sheen), was the oldest man in the village, purportedly having been born in 1898. He wasn't hard to find. Everyone in the village knew who he was and where he lived, and they were anxious to take us to see him.

We were led first to a narrow set of concrete stairs leading from the village's main road, along the riverfront, to a second flight of homes. A few of the houses seemed to be very old, little more than sticks and mud bricks. Many more, though, were newer. Albeit still quite simple, they were of wood, cement, and cinder block construction.

As we walked, a local villager told us Boxin had attained a kind of celebrity status in the region, and even throughout China. When we arrived at his modest home, though, it was clear that "celebrity" didn't come with any Western-style monetary rewards.

We climbed a small set of stairs into the second story of the basic cement home. The front room was three-sided, sort of like a dollhouse in which parts of the interior are visible to anyone from the outside. We stepped through the open space and into a small entryway. No one appeared to be home, but I heard faint sounds coming from the interior. A moment later, one of Boxin's relatives appeared outside.

The man's face contorted into what I read to be a mixture of surprise and puzzlement. As had been the case throughout our journey to this place, I sensed that my family and I might have been some of the first Caucasians these rural Chinese had ever encountered.

"Hello," I greeted him in Mandarin. "We've come here all the way from America and we wanted to see Boxin. Is he home?"

Upon hearing my Chinese, the man lit up.

"Yes, yes of course. He will be so excited to see you," the man said.

The man, who introduced himself as Boxin's grandson, told us that like most of the village elders Boxin didn't speak Mandarin, but offered to translate between my Mandarin and his grandfather's village dialect, called *Zhuǎng huà*.

We were ushered deeper inside the house, past a small partition into a kind of waiting room. The sounds from inside grew more distinct. We made our introductions to more of Boxin's family members. His great-grandchildren crowded around us, as eager to see and speak with us as I was to meet their patriarch. We were then led into a larger living room area. In a corner, to my surprise, were a few youngsters watching television; my preconception of a village where everyone is so incredibly healthy was that it would be a place where no one sat around watching TV.

One of the great-grandchildren explained that because of the number of Chinese people who wanted to meet Boxin, they'd turned this space into a kind of reception area. Along one wall sat an ornate cushioned settee, what Joshua later described as a throne. A colorful ceramic relief with mountains, trees, flying geese, and a tremendously large red Chinese hieroglyph, which I recognized as the symbol for longevity, served as a backdrop. No one occupied the central seat, but it was clear who would.

A large plaque bearing a government proclamation honoring the home's ancient owner hung as though this was some kind of museum. Along that same wall, and several others, were photographs. Almost all depicted a man with a narrow face and small, dark eyes, usually wearing a round cap. In one photo the little old man was at the center of a table with six elderly women, three on each side, smiling and conversing.

"Those are all of the centenarians together," the grandson explained.

My mind was having trouble registering what I was seeing. The people in the photograph looked as though perhaps they were in their mid-eighties.

"But this must have been very many years ago," I said.

"Not at all," the grandson replied. "That photograph was taken last year."

I looked again at the photo and four smaller ones below it depicting the same meal. All of the people in the pictures were sitting perfectly upright. Each was balancing a bowl in one hand with chopsticks deftly perched in the other. They were smiling and laughing. One was rising from her chair, stretching out to reach for something across the table.

Jane's attention was drawn to another room. She motioned to me. I stepped toward the doorway and heard a younger woman engaged in an animated conversation with someone. She seemed a bit exasperated, urging whoever she was speaking with to hurry up. A moment later, I got my first glimpse of Boxin.

At a reported age of 114, he was the oldest person I'd ever seen, and the oldest person in this village, but instead of being seated in a wheelchair or residing in a bed, he was searching intently through a closet, then under his mattress, then back to the closet again. He moved with a fluidity and intensity that surprised me, bending and stooping, turning to respond to a woman who must have been one of his great-granddaughters.

He moved like our nine-year-old son! He bent at the waist, flexed his knees, and turned his head, with the freedom and energy of someone less than half his age. I didn't hear the sort of grunting that accompanied nearly every one of my exertions.

When one of his great-granddaughters finally said, "American," the spritely old man froze. He stood fully erect, turned to look at us, and his face exploded into a wide smile. He reached out to Jane and, holding her hands in his, exclaimed, "Americans! We are friends! China and America are friends!"

"Yes!" Jane responded enthusiastically in her best Mandarin. "We are friends!"

As we walked to the reception room, I learned something remarkable from one of Boxin's great-grandsons. Even after Boxin had passed the hundred-year mark, he had continued to work in the fields, and was the extended family's main provider of food and income. Long after many of his fellow centenarians had stopped this kind of work Boxin had continued to do arduous labor.

"Only in the past two years has he slowed down a little bit," the great-grandson said.

I chuckled at that. Boxin's "slow" mode was considerably quicker than many people's "fast."

Several minutes later, Boxin returned, sporting his traditional *changshan* and black trousers.

"Come," he said. "We will eat."

I would have been perfectly content to simply sit and talk with Boxin for hours to come, but our relationship began with an invitation to share food. And that, I believe, is a very good place for relationships to begin.

I think that was the very first thing Boxin taught me. Nourishment is, after all, the beginning of everything else we do. If we're going to do something radical, such as resolving to live longer, happier, healthier lives, it should begin with what we eat.

That's where this book will begin, but this is not a diet book, especially if you think of a diet as a plan that limits the amount of food you can eat. Instead, this is a story about a village where eating good food, and *plenty* of it, is just part of a lifestyle where no one stresses out about living long, healthy, and happy lives. They just do it.

And if they can, all of us can.

LIKE ALMOST EVERYONE else in the United States, I'd tried a lot of different diets and exercise regimens over the years, without much success or benefit. I'd consulted fellow doctors and nutritionists. Everyone seemed to have a different answer for me.

But everything began to change as I came to know the villagers of Bapan.

Here was a place where people age very slowly and don't struggle with diets or obesity. It's a place where people in their nineties and even one hundreds are often still out in their gardens and farm plots, growing their own organic food. It's a place where there is virtually no heart disease or cancer. It's a place where dementia is all but unheard of. And because of these and other factors, it's a place where people have an optimistic outlook on growing old. In fact, the oldest people in the village were the most adamant that life just keeps getting better with age.

To be honest, all of this was a bit destabilizing for me. It stood in stark contrast to much of what I'd learned at Johns Hopkins and Stanford. In those places, I'd been taught that chronic medical problems were just part of aging and that we have medications and surgeries to treat these conditions. In this way of looking at life, a painful decline was pretty much inevitable; all we could do was make it more tolerable. This was also in line with the hundreds of medical studies, abstracts, and book chapters I have published over the years on cardiovascular disease, strokes, and dementia. All along I had just considered these conditions to be a normal part of the aging process.

As a cardiologist specializing in the treatment of atrial fibrillation, a condition most often brought on by our modern lifestyle, high blood pressure, and obesity, I was treating thousands of patients with that same logic. Lots of medication. Lots of procedures. Lifestyle changes that accommodated their ailments, rather than addressing the root problems.

Bapan was like a tonic to all of that.

At the time of my first visit, there were only about 550 people living in Bapan. Not surprisingly, the number of centenarians fluctuates from time to time as the eldest residents die, quickly and peacefully in their sleep, in most cases, and the relatively large number of people in their tenth decade cross the threshold into their 100s. Conservatively, though, there's usually at least 1 centenarian for every 100 people living there.

To put this into context, the average ratio of centenarians in the United States is 1 in 5,780, and that's really not bad, globally speaking. But even among the places across the globe that have come to be known as "The Blue Zones," a term coined by the Belgian demographer Michel Poulain and popularized by Dan Buettner's wonderful book about places in the world where people tend to live abnormally long lives, Bapan is off the charts. For instance, the famed island of Okinawa, Japan, quite well-known for the longevity of its residents, only has, approximately one centenarian for every 2,000 people.

There are a few caveats to this, the key one being that Bapan's oldest residents aren't likely to ever claim a Guinness World Record for longevity because there aren't any formal birth certificates to substantiate their ages. There were no censuses conducted in Bama County a hundred years ago, and China didn't begin issuing birth records until after the 1949 communist revolution. Also, with China's economic transformation beginning in 1978, most of the younger generations have migrated to the large industrial cities of China looking for work.

However, I've become quite confident that these individuals' ages are accurate, according to China's national census and their government-issued ID cards, or at least very close, for a number of reasons.

First, there is an extensive body of medical literature in both Chinese and Western medical journals on the longest-lived residents of Bama, China. Some of these studies report an age verification process, in addition to data from China's national census and their national ID cards.

Second, the living lineage—children, grandchildren, great-grandchildren, and great-great-grandchildren that I have come to know very well—supports the genealogical timeline. Many older villagers can even share their family tree and the eight Chinese characters they were given at the time of their birth, which are recorded in their family's genealogy and relate to their birthday according to the lunar calendar.

Third, the Chinese Zodiac, *Sheng Xiao*, rotates through a cycle of animals every twelve years, and every Chinese person knows his or her animal. With very rare exceptions, almost no woman gives birth at the age of twelve, and very few women naturally conceive after their mid-forties, so coupling a woman's zodiac animal with her family tree can give us even more certainty as to her age.

Fourth, the Chinese Communist Party has worked hard to identify people who participated in the Chinese Revolution, which began in the 1920s and continued through the creation of the People's Republic of China in 1949, to provide military back pay and pensions; many of this region's centenarians have records of their time in Mao Zedong's rebel army, including the age they were when they served.

Fifth, in recent years the Chinese government has started providing a monthly centenarian stipend and free medical care, including home health visits, once someone is certified as a centenarian based on all available documentation. This can be an arduous process as local government budgets are often fiscally strained. And while the stipend might seem like an incentive to exaggerate someone's age these days, there was simply no reason for illiterate people in this poor part of China to fabricate documents, potentially risking imprisonment or death under Mao's new regime back in the early 1950s, when the Chinese government started keeping track and issuing government IDs.

Finally, there are records going back hundreds of years that refer to this area of China as a place where people live to very old ages. One such record comes from an expedition launched by the Jiaqing Emperor of the Qing Dynasty, who ruled China from 1796 to 1820.

The emperor's men reportedly brought with them a birthday gift of a poem to be read to one of the area's centenarians, named Lan Xiang, who was then thought to be the oldest person in the kingdom.

On balance, I believe these people are as old as they are reported to be. But even if they weren't, it wouldn't change my perspective that there are tremendous lessons to be learned from a village with such incredibly low levels of cardiovascular disease, cancer, and dementia. That's because, in my mind, longevity in and of itself is a bonus. The truly remarkable thing is how healthy and happy these people are along the way.

IT'S ONE THING to live long, but in the time I've spent with the folks who live in Bapan, I've learned that these people don't just survive into old age; they thrive in every way. Physically. Mentally. Emotionally. Spiritually. Here, the elderly rarely need medications or surgeries and they don't hobble around or live in nursing homes. They're active, engaged members of their communities. They take walks. They work in the fields. They greet visitors. They play mahjong. They cook and clean for themselves and others. They take care of the animals and the children.

They even do kung fu! To witness, as I have, a 110-year-old woman walking down the street throwing kicks and chops is a wondrous experience.

This is what makes Bapan so fascinating, and what makes the lessons its residents can share with us so important. Because, while I was certainly interested in meeting these centenarians, I wasn't that taken by the mythical status of attaining one hundred years on the planet. To me, longevity wasn't a goal but an *indicator*. Surely, in order to grow so old, the people in Bapan must have been doing something to stay healthy, not just in their final years but *throughout their lives*. I wanted to know what it was. I wanted to know how they lived, what they ate, how they exercised, and what their environment looked like. And if they were doing something so right, maybe I could, too.

After all, most of the living centenarians in Bama County reached their eighties and nineties without ever having been to see a doctor, let alone visit a hospital. While they do have health care services available to them now, that's a relatively recent development; it has only been within the past ten years that they've had access to modern doctors. Before the late 1990s, when this village slowly became known throughout China, the average income was about 120 yuan; that's roughly $20 a year. For the vast majority of their lives, these villagers received no medical care whatsoever. Yet today they are as active and vital as people half their age, and often showing few signs of slowing down.

And it's not just how they act, it's what's going on inside their bodies, too. American twenty- to thirty-four-year-olds have a substantially higher incidence of high blood pressure than the hundred-year-olds in Bama County. Meanwhile, the rate of heart disease in the United States is 17 times higher than it is in rural China, even though there is no culture of "exercising" there. The rate of breast cancer is 10 times higher in the United States, even though there are no screening mammograms in areas of China like this. The rate of dementia is more than three times higher, and no, they don't do crossword puzzles to combat memory loss.

This isn't a case of a few random people in a remote village who happen to live longer than Americans do. This is the case of a special place in the world where health, happiness, and longevity have been a way of life for a very long time.

HOW MUCH LONGER will Longevity Village be Longevity Village? That's a very open question. There is a concern among some people in Bapan that the fundamental things that make the village so incredibly special are being inexorably changed as modern society encroaches on this little part of the world. In just the past few years, as greater China has become aware of the "miracles" that happen there, Bama County has become a vacation destination for rich Chinese seeking quick cures to their ailments. Ironically and tragically, because they're often looking for miracles, rather than wisdom, many of these "medical tourists" have brought with them their cars, their soft drinks, their cigarettes, their smartphones, their exercise habits, and their stress. An industry catering to these visitors has developed.

Bapan itself is still quite small and remote, but thousands of Chinese who had no previous roots in the area have moved into greater Bama County to participate in this "health miracle economy."

As one of the few Westerners, and even fewer Western medical researchers, to have visited before these changes really began to gain ground in the village, I feel exceptionally fortunate to have gotten to know many of Bapan's residents. We've eaten together, worked side by side, and spent countless hours talking about our different lives.

Over time, they've come to trust me as a friend, doctor, and researcher. I've studied their lives extensively. I've translated studies about them that have been published in the Chinese medical literature. My research team has even done genetic testing on many of these centenarians, and when we did, we discovered something fascinating.

It's worth noting that the residents of Longevity Village exist as an almost perfect experimental control group, quite ideal for a long-term study where results must be verifiable and extremely reliable. That's because all of the residents I've studied have lived their entire lives within the borders of Bama County. While the Han Chinese ethnic group comprises 92 percent of everyone living in China, the village is made up of people from the Zhuang Chinese minority. Intermarriage with the Hans happened during three periods of wartime when Han soldiers were sent by the emperor to guard China's southwest border. The last time this happened was around the end of the Ming Dynasty, which concluded in 1644. So, genetically speaking, this group has a rare kind of homogeneity that we seldom see in other experimental groups.

Additionally, because of the advanced age of these individuals and their history of remaining in one place for so long, variable environmental factors and influences have been kept to a minimum. In other words, the villagers were all essentially exposed to whatever was contributing to their great health and longevity in equal measure.

All too often we see diet and lifestyle studies with follow-up periods ranging anywhere from three weeks to five years, hardly long enough to really learn about long-term health benefits and consequences. Our bodies are incredibly complex machines and, like any machine, there are both short- and long-term effects to every action we take. You simply can't always infer long-term data from short-term sampling.

That's what really makes Bapan so remarkable: We can see the *lifetime* effects of diet and lifestyle choices, because everyone in the village shared those experiences over the course of many, many decades.

Now, for the most part, our genes are quite similar. If you were to pluck up any human from anywhere in the world and compare them to another randomly selected human, you'd find that their genomes are likely to be about 99.5 percent the same. But given that the human genome has something in the neighborhood of three billion nucleotides, the basic building blocks of DNA, there's a lot of room for diversity in that 0.5 percent.

Does that diversity mean that some of us are genetically destined to live long lives and others not? Not at all. Today we're learning that the impact we can have on our genes is profound. Rather than being stuck with what we've inherited from past generations, research shows the *expression* of our genes can change significantly, and positively, as a result of the decisions we make every day. That's what we've found is happening in Bama County.

Preliminary genetic work my team has conducted on six of the centenarians of Bapan has shown that the majority have genes that *should* predispose them to hypertension, atrial fibrillation, myocardial infarction, hypertriglyceridemia, and hypercholesterolemia. One of the centenarians has gene markers for an increased risk of *all five* of these conditions. Two of the centenarians have greater than a 120 percent increased risk of developing hypertension, based on what we know about how these genes typically act, yet their blood pressure is remarkably steady.

Our findings are not the exception. Other studies of people in Bama County have revealed genes that, based on everything we know about genetics, should actually *predispose* these folks to heart disease, Alzheimer's disease, high cholesterol, and diabetes. In one study, researchers found that 516 people from Bama County, all over the age of 90, carried a gene that often results in elevated homocysteine and cholesterol levels leading to heart attacks and dementia. Despite being genetically "programmed" for early heart disease and memory impairment, though, these people showed almost no signs of these diseases. In fact, studies of Bama centenarians have shown that even in those over the age of 100, heart disease is only seen in 4 percent. Another study looking at 267 long-lived Bama residents at an average age of 88 could only find one case of dementia. To put these numbers in context, about 85 percent of people over the age of 85 in the United States have already developed heart disease, and roughly half of all people in the United States age 85 or older have developed dementia.

As it turns out, the only measurable genetic difference between the people in this region who live a long time and those who don't is something called methylation, a mechanism our cells use to change the way our genes are expressed in response to how we live. And we know from studies of people from gene pools all across the world that *everyone* has the ability to positively impact their genomic expression, for good and for ill.

We've all known someone, for example, who was gifted in some way but didn't work to hone that gift, with rather predictable consequences. We've also all heard the inspiring stories of people who are naturally disadvantaged in some way, but are able to overcome that disadvantage through hard work and dedication. That's how our genes work, too. That's why, after a group of researchers considered the genetic and environmental factors impacting the lives of nearly 3,000 identical and fraternal twins from Denmark, they concluded that "longevity seems to be only *moderately* heritable." For women, the researchers concluded, only about 26 percent of longevity was the result of heredity. For men it was about 23 percent. The rest, the scientists concluded, is up to us.

When I came to recognize this, it was exceptionally freeing. For years I'd blamed my health problems on a rather poor draw in the genetic lottery, and in some ways it was true. When I had my DNA analyzed, the lab report was downright depressing. I carry variations of genes associated with obesity, diabetes, Alzheimer's disease, and inflammatory arthritis. But what the people of Longevity Village have taught me is that our genes aren't a prison sentence.

Rather than being genetically destined to live long and healthy lives, it's quite clear that the people of Bapan have benefited from lifestyle choices and attitudes that have actuated their genes in ways that have allowed them to thrive to one hundred years of age and longer without medications, surgeries, or doctor visits.

We know this, too, from looking at what happens when people leave the village and significantly alter their lifestyle. When one young man I met left the village to seek work in the city of Guangzhou, for instance, his job in a factory sweatshop and changes in eating habits took an almost immediate toll on his health. After suffering from the effects of stress, a lack of healthful physical activity, a poor diet, air pollution, and weight gain, he decided to return to the village and reintegrate, the best he could, with the traditions of his ancestors. His health was remarkably restored within months of his return.

As they say in China: *"Nǐ wúfǎ zǔzhǐ niǎo er cóng nǐ de tóudǐng fēiguò, dàn què kěyǐ zǔzhǐ niǎo er zài nǐ de tóu shàng zhù cháo"*—"You can't prevent the birds from passing over your head, but you can prevent them from making a nest in your hair." That's a quintessentially Chinese way of saying something I often tell my kids: "You can only control what you can control." And, as it turns out, we've got *a lot* of control over our genetic destiny.

The secrets of Bapan are more than simple platitudes, though. They're life lessons in health and happiness aligned quite remarkably with Western medical research.

At first, these lessons look quite simple:

1. Eat good food.
2. Master your mind-set.
3. Build your place in a positive community.
4. Be in motion.
5. Find your rhythm.
6. Make the most of your environment.
7. Proceed with purpose.

These lessons have changed my life. I'm no longer taking any of the medicines I once was. I've shed 35 pounds. My total cholesterol level has dropped from 211 to 118 and my blood pressure has dropped from 140/90 to 115/70. More important than those quantitative measurements, though, is this: I'm once again able to pursue activities that helped give my life meaning, like skiing, running, biking, and basketball. Most importantly, I'm no longer haunted by the thought that I might not be around to see my children grow up. I plan to be around to see my grandchildren and great-grandchildren, too!

These lessons have also changed my patients' lives. Hundreds of men and women, young and old, who have applied these lessons to their lives are living better, more active, and more fulfilling lives free from medications and without any procedures. Some of their successes make mine look quite unimpressive by comparison. And while I still do treat some of my patients with surgical and pharmacological interventions, when necessary, those are not the first solutions we turn to as we work together to address their health needs.

All of that is why I can be very confident in saying that these lessons can change your life, too.

Now, there's nothing inherently revolutionary about any of these ideas. But if you, like me, want to know *why* these lessons work and see *how* they've worked in other people's lives, then I'd like to take you on a journey to a place like no other in the entire world.

SOONER THAN LATER, I suspect, Bapan will be on the map. At the moment I'm penning these words, though, it's still a bit hard to find.

Given my country's military history, most Americans know where Vietnam is. If you were to start in Hanoi and head in a perfect line toward Beijing, you'd run right through Bama County, just about 200 miles into your journey.

The beautiful Panyang River cuts through the region like a sash. If you were to follow it upstream, between an endless phalanx of pyramid-shaped limestone mountains, into Bama's northwest corner, you'd be getting close.

It's an amazing place, as green as you've ever seen, owing to lots of rain, sunshine, and very few days in which frost ever touches the ground. It's also sticky hot, with summer temperatures that often reach into the 90s and an average relative humidity of 78 percent. Along a river that changes, from season to season, from jade green to chocolate brown, en route to the village, is a seemingly endless parade of water buffalo, flanked on the riverbanks by tall bamboo grass that sways ever so gently in the wind.

When you look upriver and see a mountain that looks like a massive yawning lion, you've arrived in Bapan, which is nestled between the mountain and the water at a place where the river breaks sharply to the east before returning to its generally southeasterly route.

Up until quite recently, to get to Bapan you had to walk across a rather rickety old wood-and-wire footbridge. It was often slick with morning dew. The wooden slats would sometimes rot away. The bridge would sway from side to side in the gentlest of breezes.

But there was something very special on the other side.

I

EAT

GOOD

FOOD

*"Why would I ever eat something
I do not enjoy? There is always
so much good and nourishing food
around me."* —BOXIN

ALTHOUGH HE HAD ENTERTAINED GUESTS FROM ALL OVER China, Boxin was quite interested in what had brought me to his tiny village from my home in the United States.

"I'd like to learn the secret to a long and healthy life," I explained.

"Secret? There is no secret," Boxin replied. "God let me live this long, and this is why I have lived so long."

My heart dropped. Had I come all this way only to be told that a lifetime of health was all just a matter of luck or fortune?

"Surely," I pleaded, "there is something I can learn from you about living a long and healthy life."

"Maybe there is," Boxin said. "But since I don't know what it is that is important for *your* life, I will just tell you my story."

And that's what he did, starting with details from the earliest years of his life. He was a master storyteller. He captivated me with a saga rich in details, full of emotion, and brimming with the very sort of guidance I was seeking.

Although more than a century had passed, Boxin remembered everything. He remembered the day his father died, when he was just six years old. He remembered going into the mountains to try his hand at herding animals at that very young age to help provide for his family. He remembered when each of his children, grandchildren, and great-grandchildren were born.

One of the most vivid of his memories begins in the late 1920s, when nearly all of China was descending into a brutal civil war and soldiers loyal to Mao Zedong's communist army arrived in Bapan. The soldiers were the most educated people Boxin had ever known, and he asked them to teach him to read and write. They did, a bit, but spent far more time teaching him how to use a rifle.

Boxin didn't comprehend all of what the communists were fighting for, but he and others in the village understood that, for far too long, the warlords and landlords always had enough to eat, while the farmers who produced the food often went without it.

"It wasn't fair," he said. "We were always hungry."

When we look at the problems in our life we often find that food is at the center. When we look at the problems in our world, it is often very much the same. We all need nourishment. After air and water, food is our most pressing need. And, if pressed, we will fight for it.

The Communists ultimately moved out, leaving behind nineteen guns. When the Nationalists swept back in, a few years later, the village had been marked as enemy territory. The ragtag villagers were completely outmatched by their well-armed opponents.

The Nationalists did what armies often do, leaving death and destruction in their wake. The surviving villagers of Bapan fled to the mountains. Their homes were burnt to the ground. Boxin's wife and daughter fled to her parents' home. For many months, Boxin and other villagers lived in a cave, high above the village, foraging for whatever they could find to eat. There were about a hundred of them, and there was precious little food. They ate wild vegetables, fruits, and tubers and shared everything they had in equal measure among all of the survivors.

It was several years before the Nationalists completely moved on, but eventually Boxin and the other survivors went back to the village and began to start anew.

The first thing they did was plant new gardens.

AS I LISTENED to Boxin describe the joys and tribulations of his life, one thing really jumped out at me: No matter how good or bad things were, the need for nourishment was always constant. That might seem like a rather commonsensical observation, but for me it was revolutionary. We tend to forget, and some of us really never come to know at all, all the work that goes into producing food. When we divorce ourselves from the origin of our food, it becomes easy to also divorce ourselves from the *purpose* of food.

"Food can do many things for us," Boxin told me. "It can make us happy. It can bring back memories. It can bring people together. It can also drive people apart."

But its first purpose, he stressed, is simple nourishment.

"Food is good for us!" he said. "We should eat lots of it because it is good for us!"

I'd always learned that it was important to exercise restraint at mealtime, and had tried my best, though often unsuccessfully, to do so. But Bapan's villagers ate as much as they wanted without any apparent thought as to the repercussions. That's because there actually were none.

This is true for two reasons. First, none of the calories they consume are empty calories; every bite of food was packed with vitamins, minerals, and nutrients. Second, the villagers' eating habits exist in perfect balance with the rest of their lives. And that balance begins with the way they think about food.

Making sure we make good decisions about food means keeping *nourishment* in mind. To bring this philosophy into my life, I've made a habit of reminding myself of the first purpose of food before each meal and especially before any snack. If you already say a little prayer before eating, this is an easy second ritual to add. If you don't pray (or even if you do, but find it hard to remember) putting a sign on the inside of your food cupboard and refrigerator can be an effective way to start training your brain to consider your food's purpose before you eat. Trust me: It's absolutely incredible how much power you can exert over your impulses to eat bad food if you simply devote a little conscientious thought to *why* you are eating.

A patient of mine named Kyle recently told me that he'd always questioned whether he had simply been born without willpower, and felt ashamed that he couldn't seem to control his impulse to swing into the nearest drive-thru whenever he was on the road. When he stuck a sticker on his dashboard reminding himself of the first purpose of food, though, everything seemed to change. "All those years, every time I'd get an urge for fast food, I'd told myself 'but this isn't good for me,'" he said. "Somehow that wasn't enough. But everything changed when I switched to thinking about food in a different way, not as something I can't have because it's bad for me but as something I should have because it's good for me."

The effect Kyle saw in his life has been well documented by Western researchers. When Cornell University's Food and Brand Lab looked at dozens of studies examining messages about nutrition, they found positive messages about what food can do *for* us are more motivating than negative messages about what food might do *to* us.

Of course, we can't make something like a double-bacon-cheeseburger good for us just by thinking it is. We still have to surround ourselves with the kind of food that is actually healthy for us.

———

HISTORICALLY SPEAKING, THESE are relatively good times in Longevity Village. There has been peace in this region for a half century. Food is plentiful. A steady stream of visitors, mainly Chinese from other parts of the country hoping to tap into the "magic" of Longevity Village, have been coming here since the late 2000s, bringing with them the sorts of cash that Bapan's villagers could not have dreamed about a decade ago. There is greater access to the gastronomic blessings and culinary curses of the developed world than ever before.

Yet many people in the village still eat as they ever did. A bit of animal meat, but not much. Nothing processed. No refined sugar. And, owing to the fact that most Chinese are lactose intolerant, no dairy products. Lots of roots and tubers. Lots of wild fruits, garden vegetables, nuts, seeds, and legumes. Whole grains, never refined, and not too much. In good times and bad, these foods have comprised the vast majority of their diets.

In the Western world, the word *diet* tends to evoke a lot of confusion, consternation, and even emotional turmoil. It has all sorts of negative connotations. It tends to be thought of as a set of rules.

But in Bapan there were no rules. No one was "on a diet." No one was counting calories or carbs. And with limited exceptions throughout history, they've always had enough. Poor as they were, and still largely are, it was very rare for anyone to go hungry. In fact, as Boxin told me, and as he demonstrated every time we shared a meal, the people of Bapan ate as much as they wanted, although absolutely no one was obese.

Feng Chun, who cooks and serves most of our meals when we are in Bapan, is a notorious food-pusher, and not just with the guests that stay in her family's modest inn. I love watching her with her own family at mealtime.

"*Duō chī diǎn a!*" she is always chiding. "Eat more! Eat More!"

The villagers of Bapan don't have thou-shalt and thou-shalt-not lists of foods that align with the rules and regulations of the latest fad diet. They don't have smartphone apps measuring "points." They don't have recommended daily allowance labels on the sides of food packaging. But what they do have, and what they have had for centuries, is geography. Much of their health miracle occurred because they were physically cut off from the rest of the world for millennia. While that might seem a hard thing to replicate in the modern world, in this book I will teach you how to achieve everything they enjoyed while still living a modern, connected life.

Even today, getting to the village isn't easy. Chinese health-seekers, who make pilgrimages here from other parts of China, usually have to fly into Nanning, a large regional capital known as the Green City because of the lush surrounding forests. From there, if the weather is good, it can take about four hours to make it to Bama County, and another hour to reach Bapan. There's only one road to the village, and summer monsoons can fell trees, create massive red mudslides, and turn the highway into a virtual river. When this happens, Bapan can be cut off from the rest of the world for days at a time.

Before modern roads, the trip into this thickly forested mountain area would have been infinitely more challenging, with a perilous voyage up the banks of the Panyang River being one of the only ways to reach the village. Historically what this isolation has meant is that what's grown in Bapan stays in Bapan; the villagers have been limited to eating what they can harvest. Of course, given that they live in one of the most fertile places in the world, what they can harvest is almost endless.

I was astounded by what I saw the first time I approached the bridge that leads to the village. To my left and right were thickly tangled patches of pumpkins, squash, potatoes, cabbage, and a few vegetable plants that I simply didn't recognize. Just beyond those patches were deep green geometric rice paddies stretching outward and onward as far as I could see. Along the riverbanks on either side were trees brimming with mangoes, papayas, bananas, and figs.

As I crossed the bridge I could see fishermen, ankles deep on the silty banks, pulling large nets slowly through the water. Others held long fishing poles out over the river, waiting for a bite. In a small inlet, a man was feeding a gaggle of plump white geese from a big yellow bucket. As we entered into the town, there were chickens running this way and that, some with trails of fuzzy yellow chicks behind. On the main road, just past the village entrance, someone was roasting a small pig, rubbed all over with sesame seeds, above a wood fire. Somewhere nearby, someone was cooking with peppers and garlic.

By the time I sat down for my first meal in the village, my expectations were quite high. And never in my life have such lofty expectations been completely obliterated by something *even better.*

THE SHORT, ROUND table was nearly covered with plates and bowls, each of them heaping with food. Of course there was a big steam pot of rice—a simple, unpolished varietal that, fully cooked, remained stiff but not crisp.

There was another pot of corn porridge. There was boiled pumpkin. There was a simple dish of finely shredded potatoes and carrots that had been tossed with a small bit of rice wine vinegar. There was a dish of lightly cooked onions and peppers. There were two different cooked greens, none of which I recognized at first but which I was later told were beet greens and pumpkin leaves and stalks; both had been stir-fried with garlic. There was also a rather bleak-looking gray soup with some green vegetables in it.

In that and every other meal we ate in the village, I took note of what was on the table and tried to connect it to what I knew about the nutritional benefits of certain food types. The more I studied the village diet, the more I realized that there wasn't anything particularly magical about the food in this place; they simply ate good food, and plenty of it.

Rice (and other unrefined grains)

IT'S PROBABLY NO surprise that the villagers eat a lot of rice. It's offered with every meal, without any pomp or circumstance. While more Asians are eating white rice, these days, historically it was not this way in Longevity Village. Brown rice (that's *whole* rice, including the bran and germ), is very filling while being relatively low in calories, and is a superb source of manganese, an antioxidant, which helps activate the metabolism of carbohydrates, amino acids, and cholesterol. Brown rice is also a great source of vitamin B$_6$, which helps our bodies make serotonin and norepinephrine, both of which are chemicals that help our brains communicate effectively and efficiently.

While rice is the most prevalent grain in Bapan, it's not the only choice. Corn, as you might know, isn't a native crop in Asia. It's an ancient grain native to the Americas and didn't make its way to Europe and Asia until after the Columbus expeditions that started in the 1490s. By the late 1500s, though, it was all over China and has been an important part of the Chinese diet ever since. Corn shows up in various meals in Longevity Village, but its central place in households is as part of a simple porridge—just mashed corn and water—many villagers have for breakfast each morning.

Corn is sometimes derided in the Western world, these days, largely because it's the base ingredient in high fructose corn syrup, which now accounts for up to a fifth of the calories consumed in the United States. That, of course, is absolutely unhealthy. In its natural form, though, corn is an antioxidant and a great source of fiber. That's good for gut health and helps make us feel full. In corn, as in all grains, though, the more processed it is, the worse off you are. In Bapan it's always served on the cob or freshly cut from the cob.

The other grain that often shows up on Longevity Village plates is millet, an ancient grain that might actually have a longer history in China than rice, and which is a good source of complex carbohydrates, protein, and healthy fats. Even though it is generally available in the United States and quite delicious, millet doesn't factor into a lot of common dishes in the Western world. That's a shame. I've found it's wonderful with onions, tomatoes, and spices in stuffed peppers.

Those were the main grains in the Longevity Village diet, but that's absolutely not the same thing as saying other grains are bad. There's an important distinction, though, between whole or intact grains and processed grains.

While there has been a recent flurry of books and websites demonizing grains in every form, the scientific literature does not support this belief. Indeed, in major study after study, real whole grains have consistently been associated with decreased cardiovascular mortality, cancer, and premature death. Processed grains, on the other hand, are little more than sugar, and it's clear that these products are responsible for much of the obesity crisis, as well as the epidemic of atrial fibrillation, heart failure, and diabetes that I see in my cardiology practice every day.

Dr. Dariush Mozaffarian, my friend and former classmate at Stanford, is one of the world's leading cardiologists in nutrition. When it comes to grains, he says, there are four things to consider: The fiber (more is better), the impact on blood sugar (the lower the better), the whole-grain content (the more whole or intact the better), and the structure (the less pulverized into fine flour or liquefied the better). I've found, though, that there's a simple way to summarize all of these factors: As much as possible, the grains we eat should look pretty much the way they do when they come off the plant. Indeed, that's how all of the grains served in Longevity Village look.

For those seeking to lose weight, even the healthiest of whole grains should be eaten in moderation. In Bapan, villagers ate 3 or 4 servings per day at most, far less than what the typical American eats in a day, let alone the 11 daily servings that the US government once recommended in its food pyramid.

If bread is your thing, then ideally look for one of the flourless forms with no added sugar, industrial oils, or other preservatives; these often come frozen at health food stores. Sprouted multigrain and legume-based bread, often sold as Ezekiel bread, is a good choice for people who are not gluten sensitive. Paleo bread is an option for people wishing to enjoy bread without the grains or gluten. Alternatively, you can make your own healthy bread.

What most people don't realize is that finely ground flour from grains is really nothing more than instant sugar for your body. Is it any wonder that even whole wheat bread is converted to sugar by the body faster than a Snickers bar? For me, once I was able to free myself of the addictive qualities of traditional flour-based whole wheat breads, I actually came to prefer the flourless varieties because they are more satisfying and don't leave me craving more.

Like bread, pasta is another grain-based product that a lot of people can't resist. It's also typically made of heavily processed wheat. For people who enjoy spaghetti, lasagna, or baked ziti, I recommend substituting in spaghetti squash or quinoa. You can even find great tasting pastas made from mung beans, black beans, or edamame these days. Our current favorite is edamame spaghetti.

The bottom line on grains is pretty simple: To live like they do in Longevity Village, seek to eat grains in moderation and as close to their natural state as possible.

Nuts and Seeds

THE LONGEVITY VILLAGE diet includes lots of different nuts and seeds, at least a serving or two every day. In Bapan they especially like lots of peanuts and pumpkin seeds. Yes, I realize the lowly peanut is not technically a nut, but rather a legume. Nutritionally, though, it acts like a nut, and performs like one when it comes to maintaining a healthy weight, preventing cardiovascular disease and fighting back premature death. Pumpkin seeds are a superfood. They are an amazing source of biotin, a B vitamin that helps regulate DNA formation and helps protect against ischemic heart disease. They're packed with protein, healthy fat, fiber, and magnesium, which has been connected to a significantly reduced risk of sudden cardiac death.

As I teach my patients, when you follow the seven lessons in this book it's quite unlikely that you will get hungry between meals. But if you do feel you need to snack, nuts and seeds are a great nutrient-packed choice.

Sweet Potatoes

SWEET POTATOES ARE served in Bapan for breakfast, lunch, and dinner, usually chopped and boiled to softness but not mushiness. They're relatively cheap, easy to grow, and easy to store, and are one of the best sources in the world for beta-carotene, an antioxidant, which is converted in our bodies into vitamin A and helps maintain healthy skin.

Most of the villagers eat sweet potatoes several times a week. I recommend my patients do, too. While I'm not a fan of the taste of a straight sweet potato, I don't mind them as part of a stir-fry dish. Some people love them mashed. However you enjoy eating them is fine, provided you don't undo the health benefits by what you put on them or how you cook them. (Beware, for instance, of sweet potato fries, which are often fried in unhealthy oils and covered in sugar and salt.)

Vegetables and Fruits

JUST A FEW steps outside of the village are lush hillsides rich with native fruits, berries, and, if you know what you're looking for, sumptuous root vegetables. Like bass fishing pros racing their sport boats out to their favorite spots before anyone else can get there, it's common to see villagers in this region out at first light double-timing to their secret foraging locations with an empty basket or sack dangling from their arms. Indeed, when you ask around these parts for the key to longevity, many people will tell you that, if there is such a thing, it simply *must* be the wild veggies and fruits.

"In the toughest times," Boxin told me, "that was all we ate. But even in the best of times we never stopped foraging for this kind of food. It is the best food because it is the most natural food."

Every meal in Longevity Village, including breakfast, is served with leafy green vegetables, a tremendously good source of vitamin K, which is vital to blood coagulation and helpful for strengthening bones. Despite the fact that it's really easy to get your daily dose of K (just one serving of kale or spinach will do it for you) most Americans fall short.

Other common village veggies include various varieties of bok choy (a great source of vitamins A and C), snake beans (basically a very long string bean, another good source of vitamin C and folate), and carrots (a mega-dose of beta-carotene).

As for fruits, the village bounty is a virtual Carmen Miranda hat's worth of apples (which are well-known to ward off stroke and prevent dementia), mangoes (lots of antioxidants and a good way to protect against many cancers), figs (even more antioxidants, along with vitamins A, E, and K), papayas (high in folate and fiber and vitamin C), bananas (packed with B_6 and manganese), melons (some types of which have been found to improve eye health and lower the risk of metabolic syndrome), and lychees (rich in vitamins C, manganese, and magnesium, they've long been used in Chinese medicine, in addition to being served on Chinese plates).

And then there are the peppers. While food in this region of China is bland by comparison to other parts of a country where cooks tend to push their creations toward the hotter side of the Scoville scale, just about every meal in Bapan was also served with some form of pepper. Researchers have shown that capsaicin, the active component of chili peppers that produces a burning sensation when it comes into contact with living tissue, can help increase metabolism and decrease appetite. And while hot peppers have the most capsaicin, even sweet peppers have some.

The villagers don't have a magic number for how many vegetables and fruits they eat, but having "kept score" over the course of hundreds of meals with dozens of village families, I've noticed that most meals include three or four fruits and veggies, with the latter being favored by a 3:1 ratio. I advise my patients to eat at least two vegetables and one fruit at every meal. That's *nine* servings of fruits and vegetables a day. And, as it happens, this is right in line with the American Heart Association's current recommendation.

That might feel to some people like an impossible target to hit. That's because they don't start at breakfast. The standard American diet, after all, only sometimes includes fruit and rarely includes vegetables in the morning. Most of the time, it just includes foods that are either already made of sugar or are quickly processed by our bodies into sugar. That's incredibly sad. If breakfast is the most important meal of the day, after all, then why not serve it with the most important foods? Veggies and fruits, in addition to being loaded with nutrients, are also excellent sources of dietary fiber. When we start the morning with a lot of them, we end up feeling satiated throughout the day.

One great way to get an early jump on your target of daily helpings is by turning two or three servings of vegetables and fruits into a smoothie. As an added bonus, throw in some seeds or nuts for protein and healthy fat. Healthy fat, protein, and fiber is the key to making it to lunch without hunger pains. Don't dump in sugar or fruit juice; you don't need it. If you must sweeten it a bit, try a little raw honey or a natural form of stevia. Just use some fruit, vegetables, nuts or seeds, and milk. And when it comes to milk, don't limit yourself to just cow milk. Personally, I really prefer the myriad of alternatives, including unsweetened almond, hemp, soy, and coconut.

Nobody in the village actually counts their veggies and fruits, though, and you probably shouldn't either. Instead, an easy way to make sure you're hitting the mark is to simply make sure most of your plate is covered with vegetables and some fruit at each meal.

Legumes

LEGUMES ARE ANOTHER daily staple in Bapan. As in most places in the Far East, soy is eaten daily. Soy has gotten a bad rap in the United States mainly because most Americans only eat the highly processed forms. In Asia they have traditionally eaten the whole bean in its organic state, not processed isolated compounds of soy. Indeed, many people have credited the much lower rates of cardiovascular disease, cancer, and longevity in Japan to eating "real" soy.

But soy is just the start. In Bapan, villagers eat many different beans, including mung beans, peas, and lentils of various shapes, sizes, and colors. That gives them a daily dose of a food that is high in protein, minerals, and fiber.

It's not uncommon, though, for my patients to push back at my suggestion that they add more beans to their meals. Sometimes, they just don't like the taste of beans all that much. In these cases I suggest trying them in homemade Indian food, chili, soup, or a dip like hummus. Dry roasted beans also make a great substitute for potato chips, and whole bean pasta is a lovely alternative to the processed grain variety. They often come around once they discover how delicious legumes can be when prepared right. Oftentimes, though, it's not because they don't like beans; they just don't like what beans make them do after they eat them. I absolutely sympathize. My advice for a gas-free legume experience is to try mung beans and lentils, which pack the same low-calorie, high-protein punch, but generally without giving the archaebacteria that live in our guts quite so much to snack on (and make gas out of).

If you eat at least a serving of legumes each day, you'll be eating like they do in Longevity Village.

Dairy

THERE'S CERTAINLY NO lack of water buffalo in the many villages that line the Panyang River in Bama County. Theoretically, farmers who own these amazing animals could get more bang for their buck if they kept a few cows as milkers. Water buffalo milk, after all, has more protein than milk from the Holsteins, Guernseys, and other milk cows common to Western milk production. It also has more calcium, iron, and cholesterol.

But when I asked Boxin whether he'd ever considered drinking the milk from the water buffalo, he looked at me as though I'd asked if he ever ate rocks.

"Why would I want to do that?" he answered. "I would just get sick."

There's no dairy in the traditional Longevity Village diet. Like most folks across China, the people of Bama County are largely lactose intolerant. That might not be the case for you, and if you like and can tolerate dairy, a small amount of real cheese (not processed American cheese), especially made from the milk of a grass-fed cow, which has not been pumped up on antibiotics or hormones, probably isn't going to hurt. Unfortunately, most Americans' idea of a small amount of cheese is what we put on an average slice of pizza. That might be a small amount compared to what most people in the United States eat (23 pounds of cheese each year, about the same weight in cheese as an average one-year-old child!) but it's really too much.

There do appear to be some health benefits to eating a little bit of cheese, particularly when it comes to creating a healthy gut flora, but whatever good it's doing for us is almost completely obliterated by how much bad it does to us in the quantities and types of cheeses we typically consume.

I've found that pizza is one of the toughest things for my patients (and me, too) to limit. If you love pizza, make a healthier version at home. We often add almond and coconut flour to our dough with fresh organic tomatoes, garlic, and olive oil for the sauce, sprinkled with a little organic cheese on top.

What about calcium? You don't need cheese or even cow milk for that. And despite what the US dairy industry would have you believe, you definitely don't need three servings of dairy each day for healthy bones. As it turns out, in fact, some countries with the highest intake of dairy also have the highest rates of osteoporosis. Yes, our bodies need calcium, but the idea that the best sources of it are products made of cow's milk really needs to be squashed. Calcium is also readily found in green leafy vegetables, almonds, oranges, broccoli, fish, beans, seaweed, and many other foods.

If you do drink milk, does it matter if you go full fat, low fat, or non-fat? It is not clear at this time whether fat content is important at all. If you enjoy dairy, try unsweetened yogurt or kefir with live bacterial cultures, which have additional health benefits.

Cooking Oil

THE FIRST TIME Boxin saw cooking oil in a large plastic container, he was confused. He couldn't understand how anyone would ever need that much oil for anything. "Maybe," he remembered thinking, "it was for the entire village to use for the whole year."

Until quite recently, there were absolutely no processed cooking oils in the village. What they used either came from the fish they cooked or from pressing hemp seeds, which is quite difficult to do using simple hand tools so they didn't get much.

As in the case with cheese, there are plenty of health benefits to small amounts of natural cooking oils, especially those high in monounsaturated fats such as extra-virgin olive oil. (There is also good evidence that coconut oil, once shunned by the medical community, may also have many heart health benefits.) As is also the case with cheese, though, most Americans tend to go overboard on quantity and disregard quality, choosing processed foods that have been doused in virtual oceans of industrially processed vegetable oils.

Fish

LONGEVITY VILLAGERS DO eat a fair share of fish from the river. They call these *yóuyú*, which in Chinese simply means "oily fish," and they have long been cherished as part of the village diet.

In the morning, if the river is not too high, local fishermen will walk along the muddy banks, dragging a net along behind them for ten or twenty paces. When they lift the nets from the water, the little fish, no bigger than your pinky finger, glint like diamonds in the morning sunlight. They're usually in a wok within the hour and on someone's plate just minutes later. Served whole and eaten like French fries, they taste a little like sardines, though less pungent.

The prevalence of these fish in the village diet was quite interesting to me in light of research that was being conducted by Dariush Mozaffarian. He was finding that people who had higher levels of fish oil in their diets lived longer and had a much lower risk of dying from heart disease or stroke. Again and again I was finding that emerging Western medical research was quite well aligned with traditional village practices, most of which had little to do with conscientious healthfulness. The yóuyú were simply what happened to be swimming in the Panyang River.

The easiest counterparts to yóuyú in most American markets are sardines, anchovies, and herring. If you live close to the coast, you might also be able to get your hands on some smelt from your local fishmonger. As in all other sorts of foods, wild and fresh is best, but fish that have been flash frozen aren't a bad alternative. Anything that has been canned and packed in added oils should be avoided except as a delicacy—remember it's the *natural* oils in these fish that help make them healthy for us. Water-packed fish can be a healthy alternative, too, but many are packed with a lot of salt, so be careful to watch the labels. The great thing about many of these choices, especially the smaller ones, is that they can be eaten whole or nearly whole, with skin and very small bones included, but the fins and skulls left behind. That sort of nose-to-tail eating, with an extra dose of calcium, closely resembles the way the villagers of Bapan eat.

There are plenty of other oily fish that are not quite so closely related in flavor or appearance to yóuyú, but can be a good substitute, too. To that end, I recommend the fish that are on the Super Green List of the Monterey Bay Aquarium's Seafood Watch. The fish that make this list are lowest in mercury, highest in the healthy omega-3 fats, and are sustainably caught. The list includes Atlantic mackerel, freshwater Coho salmon, Pacific sardines, and, my favorite, wild Alaskan salmon.

Other Meat

ALTHOUGH I SAW chickens running about and a few pigs being roasted, I didn't see much meat on the tables of the village elders. Did they shun it? Were they against eating it on moral or religious grounds? Did they abstain for health reasons?

"Oh no," Boxin told me. "We like meat just fine, but many of us are just not used to having very much of it." Throughout his life, Boxin said, he has eaten no more than two small portions of meat of any kind each week, and there was no animal that he considered off-limits to ending up in his bowl. In fact, there is an old saying in China: "*liǎng tiáo tuǐ de chúle rén bù chī, sìtiáo tuǐ de chúle zhuō yǐ quán chī,*" meaning the Chinese will eat "anything with two legs but a person, and anything with four legs but a chair."

The reason that people throughout Bama County have not historically eaten a lot of meat is simple and logical, Boxin said. Keeping animals was a surefire way to draw the attention of marauding armies. Given the region's turbulent history, any animals that were kept were quickly confiscated, if not by warlords and hungry soldiers then by government officials.

"The only food that could not be easily taken," he said, "is what came from the ground or the river."

For feasts and celebrations like the Chinese New Year, elder villagers do enjoy bountiful helpings of meat. When they do they eat *everything*. The muscles. The eyes. The cheeks. The brains. The intestines. Internal organs of every sort. They quite literally eat from nose to tail.

Before you pass off nose-to-tail eating as too gross to even contemplate, though, know that there are a lot of amazing chefs out there who are ready and willing to change your mind. Chris Cosentino, for instance, counts among his specialties marinated tripe, new potatoes and parsley salad, and beef heart tartare puttanesca. Tripe is low in calories, high in protein, and a great source of amino acids. Beef heart has the same benefits, and is an exceptionally good source of vitamin B$_{12}$ and iron.

If you eat meat, as about 95 percent of Americans do, it's only fair to be open-minded. After all, Cosentino notes, "meat doesn't come in little color-coded Styrofoam containers at your local supermarket." No, the avid hunter and fisherman reminds anyone who will listen, it comes from living, breathing creatures that must die for our meals.

That's why, when I brought my whole family to the village in 2015, I didn't flinch when Elizabeth, then seven years old, came running in to report that Feng Chun was about to slaughter a chicken.

"She's going to kill it!" Elizabeth proclaimed. "I think she's going to cut off its head!"

"That's how we get chicken meat," I told her.

Owing in large part to the urbanization of our world, few children these days get early exposure to the realities of where their meat comes from. But for the people in Longevity Village there is no contradiction in the simple recognition of this patent truth. The best way to honor this ancient relationship, they agree, is to appreciate *everything* it gives to us.

When confronted about their eating preferences by people who are against the consumption of animals, meat-eaters often argue that our species *evolved* to eat meat, and indeed there is good evidence that meat consumption provided the hyperdosage of protein that was a major driver of human brain development. While we may have evolved to eat meat, however, we most certainly did not evolve to eat only certain cuts of meat that we've deemed socially acceptable. And, with this in mind, if you're going to make the argument that meat is a healthy part of a human diet, wouldn't it make the most sense to actually eat more of the parts of an animal that can confer health benefits?

Offal—all those "other" parts of the animal—can be an extremely rich source of B vitamins, minerals, and proteins. And because most Americans are still squeamish about eating this way, these parts can still be purchased on the cheap, sometimes for pennies on the dollar compared to "choice" cuts of meat that might not have as many health benefits.

This is more than just a philosophy for eating. It's a good philosophy for life. Getting the very most out of what we consume, day in and day out, might just be the best way to honor the blessings we've been given. Living a nose-to-tail life means making the most out of our time, getting the most out of the clothes we purchase and products we buy, and giving others a chance to use the stuff that we don't. It means using the time we've devoted to purposeful exercise to its fullest. It means trying to identify the best in everyone.

But this all starts with food. And really, if you think about it, *everything* starts with food. For this reason, I'd like to strongly encourage you to step out of your comfort zone with your very next trip to the market. If you're a meat-eater who hasn't experimented with offal before, do so for a single meal this week. And no matter how you eat, try to find ways to integrate food into your meals before it goes bad. Wasting food, no matter its origin, isn't nose-to-tail living. With that in mind, another thing you can do right now to adopt a nose-to-tail philosophy is to go through your refrigerator today, find whatever healthy foods that would be likely to spoil if you don't do something with them soon, and work them into tonight's dinner.

The bottom line on meat: Fish, one or two times a week is a pretty good bet. And if we're going to eat red meat we should treat it as a delicacy.

If you like meat but need help limiting your intake, start thinking of it as a special treat. In deference to environmental and health concerns, a lot of people have adopted "Meatless Mondays;" one day a week in which they follow a vegetarian or vegan diet. It's certainly a couple steps further than many people are ready to go, but I very much suggest turning that movement around with "Meat *for* Mondays." Doing this on a day of the week many people loathe would give you something to look forward to, and keep your meat intake more in line with that of the people in Longevity Village.

Salt

BAPAN'S VILLAGERS DIDN'T shun salt. In fact, they really cherished it, and the reason for this is simple: It was available, but quite hard to get. "Whenever we would have salt, I would feel like a rich woman," a centenarian named Makang once told me. Her husband, after all, had to walk for five days to meet with a merchant and return with the biggest crystal of salt he could carry back to the village.

What does this mean for the Longevity Village diet? Well, it could likely be *called* low sodium, but not as low as you might think, and that's probably a good thing.

For years the American Heart Association has recommended no more than 1,500 mg of sodium per day, substantially less than one teaspoon of salt. That might not seem like such a hardship if you're only thinking in terms of the salt we sprinkle on our food, but in reality this also includes all the salt that is *already* in our food, including what is there naturally. Despite these recommendations, and years of warnings that salt is particularly bad for our blood pressure, most of us get about 3,400 mg of sodium a day.

In an effort to cut back on the salt they consume, some of my patients have put their saltshaker in the back of the spice cabinet or thrown it out entirely. What I often tell them is that it might not be a bad idea to use the shaker *more,* because at least that would mean they were likely eating real food at home. Fast foods and restaurant foods, after all, make up a tremendous portion of the salt we consume. Processed foods account for almost all of the rest. The salt we add to food when we are cooking, or even after our dinner hits the plate, usually isn't much by comparison.

While 3,400 mg of sodium is certainly on the high side, it's probably not the health hazard that we've been led to believe. New research has shown that the very small percentage of people (only about one in every 100) who actually follow the American Heart Association's guidelines for sodium not only get to enjoy bland food but may also be putting themselves at risk for other health problems, particularly if they're already struggling with diabetes, kidney disease, or cardiovascular disease.

What's the right amount of salt, then? I've pegged the traditional Longevity Village diet to about 2,300 mg a day, about a teaspoon. That's a little less than midway between what the AHA suggests and what most of us get, and I tell my patients it's a good target to aim for.

Sugar and Other Sweeteners

FRUIT SELLERS IN Bama County offer a bountiful rainbow of choices. There are apples and bananas that look much like the ones you might find in an American supermarket, though perhaps a bit smaller. There are lychees and papayas and mangoes. There are slightly less familiar fruits like mangosteen berries and dragon fruits. There are fruits that I'd never seen before or since, like a big round bunch of berries I once saw being sold that appeared like a cross between a gargantuan raspberry and a cluster of chestnuts.

With all that fruit, I expected to find a bounty of fruit juices. But as I visited people's homes, I noticed that while they often served fruits whole or cut, it was rare for anyone to crush a fruit for its juice alone.

"The fruit is so good just the way it comes," Xiu Qui, the wife of our innkeeper, explained to me one day. "Why would I make more work for myself by changing that? Why would I waste any parts that we all could be eating?"

Fruit juices are delicious and, to an extent, they can be healthy, but when we remove the juice from a fruit, we're retaining a lot of the natural sugars and leaving behind the fiber and phytonutrients that help make fruit so healthy. In fact, Harvard University researchers found fruit juice drinkers would often gain weight over time, even if they were drinking the sort of "100 percent juice" drinks that many of us associate with being a healthy choice.

Contrary to conventional wisdom, fruit juice is *not* a health drink. It is liquid sugar with fewer of the vitamins and minerals that make whole fruit so healthy. When it comes to fruit, the best bet is to eat the whole thing. If you like fruit juice as much as our family does then blend up the entire fruit in a smoothie.

It probably goes without saying that you'd be hard-pressed to find a can of soda on a Longevity Villager's dinner table. Largely because of advertising we've been exposed to since childhood, not to mention the addictive combo punch of sugar and caffeine, a lot of people have convinced themselves that their Coke-a-day habit is little more than a minor health indiscretion. In fact, just a single can of soda pop each day can put you at greater risk for diabetes, heart disease, obesity, tooth decay, and poor bone health. If there was an illegal drug that did all these things, we'd call it a menace to society. But because so many of us associate fizzy drinks with the frivolity of childhood, we treat a global health scourge like a treasured friend.

My "buddy drink" was Diet Coke. And while I never assumed it was healthy, I justified my habit by telling myself that it was *healthier* for me than a regular soft drink. After all, Diet Coke doesn't have any sugar. And that, I figured, meant it was "less bad."

Magan, another village centenarian, was the one who put "less bad" into perspective for me. "If something is bad it is bad," she said. "Even if the damage is not very much right now, it builds up over time. These are the most dangerous kinds of habits."

When we make the switch from regular sodas to so-called diet drinks, we *might* be abusing ourselves a tad less, but we're not actually doing ourselves any *good*. What's worse, because we feel as though we've taken steps toward a healthier life, we've slowed and sometimes halted progress toward the *elimination* of unhealthy consumables, which should, of course, be our ultimate goal.

Of course, all of this is assuming that diet drinks are, in fact, less bad for us. And while some would say the jury is still deliberating on that question, I'd argue we haven't even finished the trial yet. We've been studying artificial sweeteners for more than 140 years, but every year researchers discover something new. What we do know, though, is that these substances may have a similar effect on our metabolism and gut flora as high fructose corn syrup.

Such findings shouldn't really be that surprising. Most artificial sweeteners are hundreds of times sweeter than natural sugar. When we expose our bodies to such extremes, we should expect extreme reactions.

That's not just a lesson when it comes to drinks. There's simply no extra sugar in the traditional village diet. There are so many foods out there that already pack a sweet punch in their natural state. Because villagers include these foods with almost every meal, they're simply less inclined to seek out even more sugary foods.

Water

MUCH OF BAPAN'S Panyang River emanates from natural springs in the breathtaking 100 Demons Caves, just a few miles upriver. The limestone caverns, some big enough to dry-dock a battleship, were once thought to hold such evil that even during the Chinese Civil War, belligerents did not dare tread inside. But while the spirits believed to inhabit the deepest, darkest trenches of the caves are thought to be very bad, the water is very, *very* good.

There are only a few towns and villages between Bapan and the caves, with no heavy industry and little but subsistence farming upstream. As a result, the water that flows into Longevity Village is quite clean by developing world standards. On top of that, the region is blessed with seemingly endless natural springs.

Water is something many of us take for granted in the developed world. Today more than 750 million people don't have access to clean drinking water, and the majority of our planet's population lives in a state in which water is physically scarce or politically insecure. But even though water is cheap, plentiful, and generally quite clean in the Western world, few people drink as much as they should. One of the reasons is because being fully hydrated can be inconvenient. If you've ever had to cut a business meeting short, or get up more than once during a movie to use the restroom, you know this feeling.

With the exception of oxygen, though, there is nothing our bodies need more consistently than water. Water and oxygen work hand in hand to keep our bodies in stasis. When we start running low on water, our plasma levels drop, essentially making our blood thicker. As an immediate consequence of that, our red blood cells, which are essentially oxygen delivery vehicles for all of the systems in our body, can't do their jobs as efficiently.

It's rare that we let this condition get too out of control, because our body almost immediately reacts by releasing hormones that tell us that we're thirsty. For most people, though, things don't start feeling dire until our homeostatic response is already working overtime, so most of us have learned to live in a low-level state of constant thirst.

But if we simply give our body the water it is *almost always* asking for, the effects are incredibly healthful. Within 10 minutes of drinking a glass of cool water, the rate of our metabolism goes up, in no small part because our body begins working to warm the water up from the temperature that it entered our bodies to the temperature of our system, which in most people is 98.6 degrees Fahrenheit. Drinking two liters of water a day can burn as many as 96 calories; that's like taking a half-hour walk.

And while water does tend to move relatively quickly through our system, as long as we're replenishing it at a consistent rate it also takes up space in our stomachs, making us feel fuller by decreasing the release of ghrelin, a hormone associated with feelings of hunger.

All of this might be an important part of the Longevity Village secret. "Even when we were running from the Nationalists, we were never without water," Boxin once told me. "It's right there in our rivers and it drips from our caves. This is a great blessing."

Even in a time in which many US states are experiencing record drought, very few of us have reason to be concerned about the cleanliness or availability of our water. That is indeed a great blessing; and we'd do well to appreciate it by staying well hydrated.

There are no rules for how much people in the village drink; they simply take what they need. There's no sense, then, in creating an artificial target for water consumption for anyone else. Everyone needs water differently, but we all *need* water, and far more of it than most of us drink.

Instead of a target for *how much*, I'd like instead to offer you some advice for *how often*. Drink water when you're feeling thirsty. Drink water when you're feeling hungry. Drink water before every meal.

That first guideline might seem obvious, yet many of us are chronically dehydrated and don't even realize it. Our bodies are thirsty, but we're used to it, so we don't respond. If your urine is bright yellow instead of clear or very light yellow, your body is probably thirsty. If your mouth is dry, your body is thirsty. If the skin anywhere on your body is cracked or chapped, your body is probably thirsty. And if you are having a headache, your body is very likely thirsty, too.

The second guideline seems counterintuitive, but it's really not. The areas of our brain that process the sensation for hunger, and the hormones that are released in response, aren't all that different from one another. Researchers have shown that we actually have a hard time telling the difference. Often when we think we're hungry we really just need a glass of water.

Hacking the deep-brain impulse to shovel food into our mouths isn't hard. We simply need to override it with another powerful instinct: the feeling of fullness. To that end, water is a great equalizer; it can help us feel full, even when there is food in front of us that we really would love to eat, if only we could. If you've ever eaten a big dinner, then turned down an amazing-looking dessert, not because you don't want it but because you can't even fathom how it would fit in your stomach, you know how powerful the sensation of being full can be. That's why I suggest my patients eat mindfully (slow down and savor your food) and drink a glass of water about 30 minutes before every meal. That's enough time to let the water begin to signal the parts of the brain that register the feeling of satiation but not enough time for the water to pass out of one's system. I like to call this "water preloading" and it is something I do every day—and which has been demonstrated to be an effective tool for losing weight. It is also something that was done unconsciously in Longevity Village, where people often stopped by a stream or spring on the way back from work to quench their thirst before going home to prepare lunch or dinner.

And, of course, we're talking here about water. Not soda. Not energy drinks. Not coffee, tea, beer, or wine. All of those drinks have water in them, but it's not the same thing. For my patients who just don't like plain water, I recommend adding some lemon or lime, whole-fruit smoothies, vegetable juice, unsweetened herbal teas, or one of the unsweetened alternative milks.

But ideally, nine times out of ten, if a cup or glass touches your lips, it should have water in it.

Longevity Soup

MY FIRST MEAL in Bapan was gastronomically thrilling, but transcendentally disappointing. What I mean by that is there wasn't much on the table that I didn't recognize, and I was at least "passingly" familiar with the health benefits of all these foods. I'd come a long way hoping to find some kind of enlightenment, and I'd assumed that would come in the form of a culinary secret.

The only mysterious thing on the table was the milky gray soup I described earlier. But as soon as I heard its name, I was certain it was just something the villagers had come up with for out-of-towners looking for the magical elixir to long life. In fact, I later learned, "longevity soup" is a staple that has been served in Bama County for centuries, and that has been known by that name for longer than Bapan has been known as Longevity Village.

The broth is made from hemp seed, and if you're going to be so bold as to call something "longevity soup," that's a good place to start. Hemp, after all, contains all of the essential amino acids. It's a complete protein. It's packed full of healthy fat and dietary fiber. Back in the United States, a lot of confusion persists about the legality of raising or importing hemp, due to its association with marijuana. Nonetheless, you can find hemp seeds, hemp milk, hemp oil, and lots of other hemp products at health food stores and even, increasingly, at local grocers.

The greens in the soup were pumpkin leaves and stalks, which have since become one of my family's favorite foods. It takes a bit of work to strip the stalks from the outside fibers but, once you get the hang of it, it's not too hard. The leaves need no preparation at all; you can simply cook them down in a bit of water or stir-fry them in a wok or pan.

Pumpkin is another often overlooked source of vitamins and other nutrients and, as an added bonus, it's relatively easy to grow in any climate. Between the leaves and stalks (which include vitamin A and omega-3 fatty acids), the seeds (a tremendous source of protein, healthy fat, and dietary fiber), and the flesh (a mega-helping of vitamin A and great source of vitamin C), pumpkin is a veritable superfood, which means Longevity Soup was essentially a concoction made of two superfoods.

Longevity soup is really quite easy to make in a modern kitchen. Here's how it's done:

————————

1 cup raw shelled hemp seeds

1 cup pumpkin greens (narrow stalks about as thick as a
*   pencil work best)*

3 cups water

½ tsp salt

To prepare the greens, first rinse and dry. Remove all the stalks larger than the diameter of a pencil. Clip off the leaves and peel any thick fibers from the smaller stalks. Other greens, such as spinach or tuscan kale, are delightful alternatives.

To prepare the broth, put the raw shelled hemp seeds and water in a blender and blend until smooth. Transfer the broth to a pot. Add the greens and bring to a boil for about 1–2 minutes until greens are softened. Remove from heat and salt lightly to taste.

————————

How often should you eat longevity soup? That depends a lot on whether you like it (I do but I know not everyone shares my assessment) and whether you've also got a healthy pattern of consuming amino acids (like watercress, chia seeds, and leafy greens); proteins (green peas and quinoa are great sources); and healthy fats (like those found in nuts and avocados).

Some of the villagers eat longevity soup every day. Others eat it only once in a while. If you can work it into your diet once a month or more you'll be better for it. But that's really just the beginning, of course.

The real secret of the soup, if there is one, is simplicity. And that's a lesson that extends not just to food, but everything else in life. The more complicated we make things, the less healthy for us they become.

WHEN I FIRST began thinking about how to explain the Longevity Village way of eating to my patients, and how to help them translate it into their own lives, I came upon the idea of calling it "the Real Food First Diet." After a while, I realized that the very word *diet* was so steeped in anguish for many people that it wasn't a productive way to describe a philosophy of eating.

For far too long, people have dreaded taking on the challenge of a new diet because they are afraid that they're going to be hungry. And, not surprisingly, when they're on a diet they often *are* hungry. Moreover, most people consider diets to be something that they "go on" and "go off." That's nothing more than a recipe for failure.

What I recommend for my patients is a way of eating that is sustainable for the rest of their lives, does not require them to be hungry or to count calories, and works in harmony with their genomes, hormones, and metabolisms. To that end, I want you to eat. I want you to eat some more. I want you to eat as much as you feel you need and I don't ever want you to think of food as something your body feels like it needs but isn't allowed to have.

In Bapan, a local fisherman in his mid-thirties named Buxiao once told me, "We eat and eat, and we do not stop until our bellies are full." These days, he noted, there's a lot greater access in the village to sugar and artificially sweetened foods, "and it is true that I have enjoyed those foods when I have had them in the past, but I never want them because I am always full of the good food nature has given to us."

If you dine as they do in Bapan, eating vegetables, fruit, a healthy protein, and a healthy fat before anything else, your body is incredibly unlikely to tell you that it's still hungry. Ergo, you're *really* unlikely to want to supplement your calories with a soda, dessert, or anything else that is bad for your daily and long-term health.

Eating healthy isn't restrictive. A standard American diet is the exact opposite, because the only way to eat the sorts of food most Americans eat on a day-to-day basis without doing tremendous damage to your body is to severely restrict your intake to the point that it's no fun whatsoever. Let's face it: Who wants to eat just one bite of a double bacon cheeseburger, eat just one French fry, and take just one sip of Coke?

Same deal for snacking. If you're hungry and you feel like your body needs something to keep going, don't battle hunger; that will just cause you to make poorer decisions about food and other activities later on. The feeling of hunger, after all, is our body's way of signaling that we are lacking critical nutrients. For those who want to snack, I encourage you to ask yourself whether you are hungry or looking for a distraction. If all you feel like is junk food, perhaps you are looking for a distraction. In this case, give yourself permission to make a different and mindful choice that may not even involve food. If you are truly hungry, you will likely feel like eating nuts, seeds, legumes, veggies, or fruit. I find that eating these healthy foods when hungry usually satisfies my hunger and I am able to forgo the junk food.

Almost all of my patients who have taken on the challenge of eating this way have been successful because, as a patient named Miguel noted, "it's hard after eating all those healthy foods to even look at something unhealthy. When I'm satiated, I just don't crave junk. In fact it makes me a little sick to my stomach to even think about."

Why does this work so well? Quite simply it is because all calories are not created equal.

Take boxed cereals for example. Despite what the packaging might claim, these factory-crafted substances, in which the most prevalent ingredient is often sugar, aren't at all a "part of a complete breakfast." Even if sugar is not found high up in the ingredient list, given how highly processed the grains in cereals are, the body will quickly convert it to sugar. And while the nutritional label might promise a low-calorie meal, that's a tremendously poor indicator of healthfulness.

When you eat a 110-calorie serving of a sugary cereal, you get an almost instantaneous spike in your blood sugar, which in turn signals the beta cells in your pancreas to produce insulin, which in turn signals your liver to make fat. Meanwhile, the sugar activates the reward center of your brain, releasing dopamine into your system and essentially giving you a quick "high." Once that cycle is complete, the glucose levels in your blood crash, and that activates a hormonal response that tells your brain you're hungry all over again.

Contrast that to what happens in your body when you eat a serving of almonds. Firstly, the one-two-three punch of protein, fiber, and fat in a serving of nuts fills you up. And because it takes a little extra work to digest, the effect of eating the almonds actually increases your metabolism. As your body goes to work slowly and deliberately breaking down what you've eaten, you get a healthy dose of vitamins B_2, E, and H—the latter of which, also known as biotin, helps metabolize fats. All these vitamins and a bunch of additional minerals, by the way, exist *naturally* in almonds; they weren't pumped into the nut for the sake of convincing consumers that they are eating healthy when they're clearly not.

Because there is very little natural sugar in an almond, and it is digested very slowly, there's almost no effect on your blood sugar, which means there's no trigger for the release of insulin, which means your liver is left alone to do more important things than making fat. There's also no dopamine response, so if you find yourself feeling happy while eating a handful of almonds, that's probably because you know you're doing something good for your body, not because you're experiencing a sugar high.

What can you do to live a calorie-agnostic life? Grab a Sharpie marker and go through your refrigerator and cupboard, blacking out the calories and circling the ingredients. Ingredients are a far more important indicator of healthfulness. And, as the food journalist Michael Pollan often advises, if you don't recognize one of the ingredients (or can't pronounce it) don't eat it.

WE'RE EVOLUTIONARILY PROGRAMMED to enjoy sweet foods. And, left strictly to the whims of evolution, we'd probably be just fine running around the wilderness eating all the wild fruits we want. But something happened in our species' development that threw a monkey wrench in the works: We figured out how to make sweeter and sweeter foods and to work less and less hard to get them. As a result, the sort of sweetness that many of us have come to expect from many foods is completely out of line with what our bodies actually need.

Processed foods and fast foods have been manipulated by food scientists to activate the reward centers of our brain. That hit of dopamine and opiate-like reaction many of us get every time we bite into a Froot Loop, Oreo, or a doughnut drives us to overeat even when we are not hungry, a fact that was driven home for me in a conversation with a patient named Mark a few years ago.

"Don't you ever get hungry?" I asked him after he quickly dropped 70 pounds following what was supposed to be a "routine" neck-fusion surgery.

"Definitely," he replied. "The only difference now is that food just tastes like cardboard."

Unfortunately, during Mark's operation, the surgeon had accidently "nicked" a branch of the laryngeal nerve, leaving him unable to taste food. Although he still got hungry, he had no way to taste his food, so the reward centers of his brain could no longer be activated by "hyperpalatable" fake foods. Thus, after battling with his weight for most of his adult life, Mark almost immediately dropped from 230 to 160 pounds.

I'm quite sure Mark would do anything to be able to taste food again, but the advantage to his situation was that his brain was no longer conspiring with his taste buds to try to get him to consume foods that have been specifically designed to get him to eat more. The rest of us, though, have a very big fight on our hands against these types of foods.

How big of a fight? Consider this: One clever undergraduate research project at Connecticut College demonstrated that lab rats enticed to choose between Oreo cookies or rice cakes preferred the Oreos about the same amount of time as rats choose to get a shot of cocaine when given a choice between that powerfully addictive drug and simple saline. The students also measured the rats' expression of a protein associated with pleasure and addiction, and found that the Oreo-munching rats had a greater activation of this protein than those who were "riding the white pony."

Even if you're skeptical of that finding, and there's fair reason to be, you almost certainly realize these sorts of foods aren't good for you. One big problem, though, is that our society tells us food like this can be eaten in moderation. One key lesson we can take away from the Longevity Village elders, though, is that they *never* ate food like this. Of course, that's because they didn't have access to it. The best way, then, to protect ourselves from junk food is by limiting the ease at which we have access to it. We'll talk a lot more about this idea, but the most important takeaway, for now, is not bringing junk food into our homes; once it's there, the battle for moderation is already lost.

It's perfectly natural to like foods that are sweet. And, as long as they are real whole foods, you are probably fine. But we must train ourselves to recognize sweetness for what it generally is in today's processed foods. Sweetness, food cravings, or eating when we are not really hungry is a signal. It's a warning sign that what we think we want might not be in line with our best interests.

In the meantime, simply eat real food first. Eat a lot of it. Eat until you are full, and then some. That's what they do in Longevity Village.

I WAS TAKING a walk across the main village bridge when I came upon our cook, Feng Chun, who was loading a basket with greens from her family plot.

"Beet greens!" she told me, "Aren't they beautiful?"

"They are," I replied, marveling at the bright green leaves and their deep purple veins. "What will we have with them?"

"I'm not sure yet," she said. "Let's walk down and see what the fishermen have gotten today."

A quick trip to the water's edge revealed the morning catch hadn't been very good; prices would undoubtedly be high. Having seen the fish, though, Feng Chun decided that's what she wanted to serve, so she sent her husband out with a net to catch our lunch. A few hours later, he returned with a half-kilo's worth, many still flipping and flopping about in his basket.

Villagers regularly eat for lunch vegetables that were harvested in the morning, then have for dinner what was harvested after lunch. The beauty of such an arrangement is that it reduces farm-to-plate time to almost nothing, and that has two major benefits. First, fresh food simply tastes better. Second, it's better for us.

A lot of people don't even realize how good vegetables, fruits, roots, legumes, and meats can be because their idea of "fresh" is something that was purchased from the grocery store. Sure, that's probably fresher than something that was cooked, canned, pre-served, and stored in a can in your pantry, but it's not really *fresh*.

It can take as long as five days for a vegetable harvested on a farm to arrive at a distribution center. From there it can be another two days to get to a grocery store. At the store, a veggie might spend two or three days in the refrigerator or on the shelf before you pick it up. And so, even if you cook it that evening, it might be a week and a half old before it hits your plate.

You can appreciate how that time and distance matters for the taste of our food if you've ever had a tomato straight off the vine in your own garden. Sweet and juicy. Firm and a little salty. None of that mushiness that we all too often have come to associate with tomatoes bought at the grocery store (usually because they were picked too early and expected to ripen in transit).

Fresh vegetables and fruits are *better for us*, too. That's because many lose significant nutrients in the days after they are picked through a process called respiration. A green bean, for instance, will lose 77 percent of its vitamin C within a week of being picked.

Even more troubling: Some vegetables and fruits are less healthful to begin with, because they've been grown in nutrient-deficient soil and bred with size, growth rate, pest-resistance, and shelf life in mind, rather than nutrition. One study from the University of Texas revealed declines in the vitamins and minerals found in 43 different vegetables and fruits compared to the same products grown 50 years earlier.

What's the answer to this conundrum? Well, there are many, but the one I like is to do what the best chefs in the country do: Buy produce grown as close to you as possible and know as much as you can about the conditions under which it was grown. And the best way to do that is to actually be able to meet and talk to the people who are doing the growing. With this in mind, I encourage my patients to shop at farmers' markets as often as possible to cut down on farm-to-plate time and create opportunities to understand more about their food.

While there are thousands of farmers' markets across the United

States, though, there are still far more communities without that sort of resource. The US Census Bureau estimates that about 80 percent of Americans live in urban areas and need to travel more than 10 miles to get to an urban farmers' market.

Another way to address this problem is to join the modern "victory garden" revolution. More than a third of Americans, some 42 million households, are now growing food at home or in a community garden. Millennials are the fastest-growing segment of the gardening generation, and families with children are especially likely to participate in growing food at or near their homes.

Now, a lot of people believe they don't live in the right climate for a garden. The truth is, though, that there are very few places in the world where the conditions are so severely inhospitable that you can't grow at least a bit of your own food in a backyard, roof, or balcony garden, especially if you're willing to build a small greenhouse. If soldiers in Iraq and Afghanistan can do it (they did) and researchers in Antarctica can do it (they have, too) then there's not much excuse for those of us who live in much less extreme conditions. Whether you live in a compact apartment in a bustling city or a sprawling mansion on acres upon acres of land, you can grow a portion of the vegetables and fruits you need to support your health.

Every window in your home is a potential greenhouse in which you can grow carrots, microgreens, salad greens, and tomatoes. Your basement is a perfect mushroom farm. Your roof is a vastly unutilized surface for planter boxes of all sorts; I've been especially encouraged, over recent years, to see my city-dwelling patients band together with others in their buildings to get access to their rooftops for the purposes of building cooperative gardens. If you live in a home with a front yard and have ever kept a lawn then you can keep a garden, and once you've gotten it started, it can actually take less water and less time to tend.

A lot of people I know only get to enjoy the bounty of their garden for a few months out of the year, generally the late summer and early fall. That's in part because they're not planting seasonal crops, but also because they don't grow crops that store well without the added effort of canning or energy costly freezing. As it turns out, some of the crops that last longest, with really minimal effort, are mainstays in the Longevity Village diet. Proper curing (a simple process of keeping root vegetables and squash relatively warm for a couple of weeks before long-term storage) results in sweet potatoes and pumpkins that can be stored in a cool, dark, dry place for several seasons.

If you don't have something edible growing in or around your home already, it's time to start. All it takes to make a small but meaningful step toward better health is a single plant. Over time your garden will only get bigger, better, and easier.

Regardless of where your vegetables and fruits come from, though, the simple fact is that these are the very best kinds of food for us. Studies have repeatedly shown that, in general, the more vegetables and fruits you eat, the more likely you are to lose weight over time. Most of us can't perfect the conditions under which we receive our vegetables, but we can eat more vegetables. That's the best place to start.

What's the right approach to ensuring a consistent, healthy supply of truly fresh food for you and your family? That depends a lot on where you live, what kind of home you have, how you work, what your transportation options are, and how willing you are to get down in the dirt. For most of my patients, the solution is a combination of food-sourcing approaches that include gardening, farmers' markets, traditional supermarkets, and eating out at restaurants that offer healthy menu options.

A CENTENARIAN NAMED Mawen told me she was deeply troubled when prepackaged, processed foods began arriving in the village a few years ago.

"Everyone told me, 'Grandmother, our leaders would not allow these foods to be sold if they were not healthy,'" she told me. "But I was suspicious. I do not know why."

Mawen didn't know anything about the dangers of lab-created hyperpalatable foods, but her instincts about processed provisions couldn't have been better. There's really no way to do processed foods right. These foods are high in sugar, salt, and unhealthy fats, and often loaded with chemicals and preservatives. In fact, many processed foods might have a longer shelf life than we do.

Despite this, processed food makers make a lot of health claims on their packaging. Increasingly, food companies are giving their products names with words like "health" "Earth" and "natural" and packaging them with brown and muted green tones. Don't be fooled. Processed foods are not health foods, no matter how they are marketed.

When was the last time you saw a health claim on a head of broccoli or a bunch of carrots? Real food doesn't need a label telling us that is it good. Any packaged food with a health claim on the package is probably a good sign that it actually *isn't* all that healthy for you.

Reading ingredients is a good start, but a better bet is *buying* ingredients. You know what a loaf of "healthy" whole wheat bread from many natural foods markets has in it? Well, very finely ground wheat flour, for starters. If the sugar high from powder-like wheat flour is not enough, food manufacturers often add in multiple other forms of sugar like juice, honey, and molasses. After sugar, there are often industrial processed oils and extra gluten (as if there was not enough gluten in the whole wheat alone). With the exception of the gluten itself, food companies do the same sorts of things to the typical gluten-free breads. We don't need any of that stuff in our bread.

We often choose processed food on the notion that it saves us time. It really doesn't. And all it takes to recognize this is to simply purchase one item, in its freshest form, that you might otherwise purchase prepackaged, no matter how fresh it might seem in its packaged form.

For instance, I used to buy a lot of bagged carrots, reasoning that the peeling, chopping, and cooking time that I'd save was worth a little more money. As it turns out, though, these so-called "baby" carrots are nothing more than bigger carrots that have been shaved down in the processing plant, thus discarding the healthiest outer layer. Moreover, even the organic varieties are often treated with chlorine. That's true for other processed produce, too, like "washed and ready" lettuce packages.

When I forced myself to buy fresh carrots, instead, and took them home to make dinner, I found that I could wash them, cut them into strips, and get them onto a plate in less than two minutes. They're fresher, they taste better and, in reality, they take almost no additional time.

Carrots are a simple place to start. Lettuce is another. Making your own hummus, salsa, or kefir are excellent next steps. And homemade, whole-grain bread, fresh whole grains that you actually grind yourself to avoid the finely ground store-bought flour, is a great goal that is also extremely achievable; once you've mastered your favorite recipe, it takes just minutes in the morning to ensure you'll come home to the amazing smell of a warm, fresh loaf in the bread maker.

Take it one food at a time. Each small step we take to assure greater freshness and health makes the next step that much easier, and many of my patients have been quite pleasantly surprised by how much their ideas about "instant" food can change in just a few weeks.

NO ONE TAKES vitamins or supplements in Longevity Village. Villagers obtain all of the nutrients their bodies need by simply eating the real food that is all around them and living life as they always have.

That's my recommendation for most of my patients, too. Need more omega-3s? Eat fish or have some walnuts. Want more vitamin D? A serving of salmon will do that for you. For your selenium have a Brazil nut each day. If it is iodine you lack then eat kelp. Let *food* be your vitamins and supplements.

I've had patients come to me who are taking twenty or more different supplements, and many of them have no idea what the side effects are, or how the supplements interact with each other. They simply heard these were good to take, and so they took them.

Even if they did have a good understanding of what those supplements are supposed to do, they might be getting less (or more) than they bargained for. In early 2015, the New York Attorney General's Office announced the results of an investigation that revealed only one in five supplements from national chains such as Target, Walmart, GNC, and Walgreens contained what their labels actually said. That's right: 80 percent of the time, according to the prosecutors, these supplements were mislabeled. But even if it was 5 percent, would you be comfortable with that? For me the bigger issue is that it is hard to know what is actually in your supplement bottle.

Is there a time and a place for vitamins and supplements? For some people in some situations, yes. But a supplement should be just that—something that you can't get any other way. According to research to which I contributed with cardiologists and nutrition experts in Utah, for instance, most people in my home state are vitamin D deficient, as that vitamin is hard to come by from food alone during our long, cold winter months. The same was found to be true by researchers from the State University of New York at Buffalo, who have observed that residents of snowy, northern US cities are often vitamin D deficient, with the elderly, pregnant and nursing women, and people with darker skin tones at particular risk. For people in these situations, and particularly those who already suffer from conditions that include vitamin D deficiency, a supplement might be in order.

First and foremost, though, it behooves us to try to get all of our vitamin and supplement needs from eating real food. In the rare cases in which that isn't enough, only *then* is it time to work with your health care provider to see if there are any specific vitamins or supplements you may need for optimal health based on your lifestyle, where you live, and your genome.

BOXIN WAS AMONG many of the villagers from every generation who coupled the traditional way of eating with a habit of going without a morning meal.

"It's good to be a little hungry sometimes," he told me, "it makes you remember what you are working for."

Eating after sunset was also extremely rare. Electricity, after all, is a relatively recent development in the village, and so historically there was very little reason to stay up much past sunset. The effect of this custom is that most villagers would have a period of at least 12 hours each day in which they would not eat, and those who eschewed breakfast would regularly go upwards of 16 hours without food.

I came to learn this at the same time another friend and colleague at my hospital, Dr. Ben Horne, was leading a number of research studies that investigated the role of intermittent fasting on human health. His work has shown that people who occasionally go without food are less likely to suffer from many health conditions, including coronary artery disease and diabetes. Animal studies have shown that intermittent fasting can increase longevity by as much as 40 percent.

In part, the health benefits of intermittent fasting are the reasonable result of giving our digestive system, and the rest of our body, a chance to reset. From the moment a piece of food touches our lips, after all, nearly every system in our bodies launches into action, like a forensics investigation team working to solve a mystery, identifying what the food is, what it's made of, and how to digest it. Hormones go this way and that, telling various organs how to respond. Those organs, meanwhile, launch into action. And, of course, we generally don't just eat one type of food at a time, so mealtime in the human body is a lot like Grand Central Station at rush hour.

None of this happens in a void, though. When our bodies are focused on figuring out what to do with all that food, it's more difficult to address the myriad of other things that our bodies need to be doing to remain healthy. Taking a break from food gives us a chance to address these other needs.

It doesn't stop there. Intermittent fasting has been shown to change the expression of many genes. One such effect appears to optimize levels of insulin growth factor, which plays a role in suppressing cancer and diabetes.

Christians, Jews, and Muslims have practiced intermittent fasting for millennia. Growing up in a religious home, I was taught from a young age that fasting for 24 hours once each month was a way to gain greater spiritual awareness. Unfortunately, the hunger pain of fasting for 24 hours was usually too intense for me to obtain much spiritual benefit. Recent reports, though, have suggested that fasting for shorter periods of time, even just 12 hours, might confer the same health benefits as a 24-hour fast.

I didn't meet any villagers in Bapan who said they regularly went an entire day without food, though some had gone that long or longer on various occasions throughout their lives through no choice of their own. Most, however, hit that 12-hour mark on a very regular basis.

To me, that was quite promising. Like most people, I was already going at least 7 hours each night without eating when I was in bed. By simply eating dinner at a reasonable hour, and refusing the temptation to snack before bed, I could hit 12 hours on most days with little effort, and have now for years.

Since that time, many of my patients have shared with me that this approach has been a significant factor in their successful weight loss, which of course leads to all other sorts of health benefits. Most of these patients now seek to limit their "feeding window" each day to 12 hours; some even compress that time to 6 or 8 hours, and the effects have been little short of amazing. And vitally, from a psychological perspective, they have told me that they no longer fear "hunger." Conquering the fear of hunger has allowed them to eat when they are hungry rather than allowing this fear to cause snacking all day long.

For my patients, though, I recommend no snacking before bed thereby giving them a 12-hour window in which they give their bodies a chance to reset by refraining from eating. And I'm fully convinced that those who can work this Longevity Village practice into their *daily* routine are better for it.

I OFTEN ASK my patients, and especially the participants in the support groups I lead, to keep track of the foods they're eating. Here's what Miguel recently submitted:

SUNDAY BREAKFAST: Almond butter on flourless bread; blueberry, kale, yogurt, and walnut smoothie; water

SUNDAY LUNCH: Brown rice and lentils with steamed broccoli, zucchini, and cabbage (with soy sauce); orange and almond milk smoothie; water

SUNDAY DINNER: Broiled salmon with lemons; mashed sweet potatoes; corn on the cob; spinach with red peppers and garlic; white wine; water

MONDAY BREAKFAST: Spinach salad with blueberries and walnuts; almond milk; water

MONDAY LUNCH: Quinoa and lentils with steamed broccoli, carrots, and bok choy (with soy sauce); water

MONDAY DINNER: Portobello mushroom sandwich with red leaf lettuce, tomato, green peppers, pickles, and onions (with French mustard) on toasted whole-grain flourless bread; corn on the cob with garlic and paprika; apple slices; coconut milk

TUESDAY BREAKFAST: Steel cut oatmeal with coconut milk; blueberry, kale, yogurt, and almond smoothie

TUESDAY LUNCH: Tomato, spinach, zucchini, and carrot flatbread wrap (with French mustard); peanuts, apple slices; water

TUESDAY DINNER: Flatbread wrap with crushed sardines, cucumber, jalapeño, green bell pepper, and onion (with vinaigrette dressing); local hefeweizen; water

WEDNESDAY BREAKFAST: Hominy porridge with coconut milk and nutmeg; kale chips; banana and an orange; water

WEDNESDAY LUNCH: Brown rice and lentils with steamed broccoli, zucchini, celery, and bok choy (with soy sauce); orange and almond milk smoothie

WEDNESDAY DINNER: Brown rice with organic tofu and honey-glazed carrots; spinach with red peppers and garlic; mashed sweet potatoes; white wine; water

THURSDAY BREAKFAST: Poached egg, avocado, and balsamic vinegar on whole-grain flourless toast; coconut, banana, almond smoothie

THURSDAY LUNCH: Brown rice with steamed bok choy, celery, red bell peppers, and garlic (with soy sauce); strawberry and banana smoothie; water

THURSDAY DINNER: Teriyaki poached salmon; brown rice and lentils; carrots, celery, and red bell peppers; coconut milk; water

FRIDAY BREAKFAST: Fruit bowl (pineapple, apples, pears, honeydew, watermelon); coffee (x2)

FRIDAY LUNCH: Tomato, spinach, zucchini, and green bell peppers flatbread wrap (with French mustard and olive oil); kale chips; orange and almond milk smoothie

FRIDAY DINNER: Grilled lean chicken; carrots; summer squash; black beans with diced red bell peppers and green chiles; local pilsner; water

SATURDAY BRUNCH: Brown rice with coconut milk, blueberries and slivered almonds; sweet potato hash; rainbow chard, spinach, green onion, and cheddar omelette; water

SATURDAY SNACK: Almonds and water

SATURDAY DINNER: Longevity soup; portobello mushroom sandwich with spinach and roasted red peppers (with mayonnaise), kale chips; corn on the cob; water

I really couldn't have been prouder. That's about as close to a Longevity Village diet as anyone is likely to get in the Western world.

I know from my conversations with Miguel that dinner was over each night by 7:00 p.m. and breakfast started at 7:00 a.m., meaning he had 12 hours each day of fasting time, with an even longer period between Friday's dinner and Saturday's brunch.

Miguel had just turned forty-two years old, weighed 245 pounds, and had an abnormal heart rhythm and troublingly high blood pressure when he started implementing the Longevity Village principles to his life. Over the following year, his transformation was absolutely inspiring. By the time he reached his forty-third birthday his weight was down more than 40 pounds, his heart rhythm was stable and his blood pressure was down to levels that would be considered healthy for a man *half* his age.

"For years I thought that the only thing worse than being so out of shape was eating 'healthy' food," he recently wrote to me. "I didn't know how much I was missing. I used to eat hamburgers and French fries four or five times a week. That tasted good to me. Now I realize that I was missing out on so much great food. I love this way of eating and I'm planning to eat this way for a very, very, very, long time to come."

Miguel's food choices weren't the only change he made, but they have been a big part of his overall success. But even though his weekly menu is a great one, it doesn't have to be *your* menu. Miguel's way is just one of many mindful approaches to healthy eating that is part of a Longevity Village lifestyle.

OFTENTIMES, WHEN RESEARCHERS look at communities where people are living longer, healthier lives, they tend to focus on what is similar about their diets. In *The Blue Zones,* Dan Buettner noted that beans, whole grains, and garden vegetables were staples in nearly every community known for its longevity. Legumes, in particular, have been shown to be a vital dietary factor in longevity. Studies show that even increasing your intake of beans just by 20 grams a day (that's really just a few beans), might decrease mortality by 7 or 8 percent. All of that holds true in Bapan.

What we tend to ignore, though, are all of the differences that exist in abnormally long-lived communities. Some of these communities eat lots of cheese, which was recently shown to have some benefits for metabolism and possibly reduce obesity. Some eat no cheese, which according to the Physician's Committee for Responsible Medicine is a good thing, since cheese is so high in saturated fat, sodium, and cholesterol.

Cheese or no cheese? Meat or no meat? Kosher or Halal? Paleo, vegetarian, or Atkins? What should *you* do? Well, you should trust yourself. Your genes are different from anyone else's. Your diet probably should be, too.

The truth is that across time and around the world people have thrived on many different diets. Just about the only diet that *doesn't* seem to work for anyone is the Standard American Diet—the acronym for which, SAD, is as appropriate as they come.

If you're working hard to minimize sugar, eat lots of vegetables, and get rid of processed foods, then you're probably doing a good job for your health and the health of your family, regardless of your other eating habits. And if you're struggling with that, as just about all of us have at times, then you're not alone. That's what the rest of this book is for.

The important thing to know is that there is no quick fix. I had to learn that lesson, too. When I first came to Longevity Village, I believed that I would be able to hone in on a dietary secret that would lead to a longer, healthier, happier life. And it's true that I learned a lot about food and our relationship with food during that first trip. But during that stay, and in all my time in Bapan since then, I've learned so much more.

And perhaps the most important thing I've learned is that the best thing we can do for our health is to make sure that eating isn't a vice, but rather a virtue.

CONFUCIUS ONCE SAID, we should "not, even for the space of a single meal, act contrary to virtue." When I learned this I was devastated. I figured it was impossible. I might eat lots of veggies, fruits, and legumes. I might never again have another Diet Coke. But what if my friends were to invite me out to pizza? What if I wanted to share an ice cream sundae with my daughter? Would that be acting contrary to virtue?

When I asked Boxin what he thought of Confucius' instructions, he gazed out of his open parlor and paused contemplatively for several seconds.

"Virtue isn't just one thing we do," he said. "It's everything we do."

Virtue, I've since come to believe, isn't just what we eat. It's how we eat. It's who we eat it with. It's our relationship with where our food comes from. It's the decisions we make about how to prepare it. It's our determination to honor the energy it gives us in positive ways.

When I came to this realization, another quickly struck me: If I was going to change my life, I wasn't just going to have to master my appetite. I was going to have to change the way I was *thinking* about my life. I was going to have to master my mind-set.

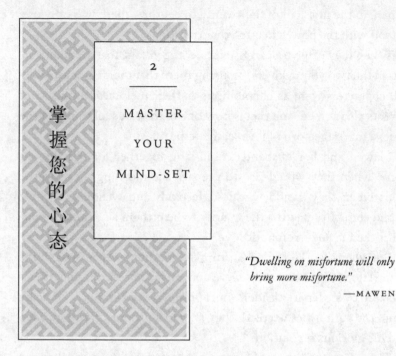

*"Dwelling on misfortune will only
bring more misfortune."*

—MAWEN

THE SINGING CONTESTS WOULD LAST FOR DAYS, BUT THERE WAS never any question who would ultimately win. Mawen had the most beautiful voice of all the women in Bama County and, when she was approached in her early twenties by a young tenor who held an equally commanding ascendancy over any of the local male singers, it seemed only natural they should team up. Their unrehearsed, improvised love songs were an exuberant back-and-forth banter about work, play, and reciprocal affection. The songs are still spoken of today, even though it has been a very long time since anyone has heard the singers.

The duo had been singing together for several years before they realized they were no longer simply *telling* stories. They were *actually* in love, a condition that could be heartbreakingly problematic in those times. For thousands of years, after all, many Chinese families had maintained an exacting tradition of arranged marriage. And in Bapan, in the first half of the twentieth century, there was only one person with the power to buck that tradition.

When they came to learn their fate, the village matchmaker surprised the two young lovers, assuring them that their eight astrological characters were as harmonious as their melodies. Both sets of parents also agreed, and that is how one of the rarest of things at the time, a marriage born of love, came to be.

Mawen and her husband would sing together for decades. At home when they would rise to prepare the morning meal. In the fields when they would work. At festivals and weddings. People would come from across the county to hear them sing, and to marvel at their *diqíng*—romantic love.

But when Mao Tse-tung's Cultural Revolution swept across China in the late 1960s, their voices were abruptly silenced. Singing, it was declared by local party leaders, hurt productivity. If songs were to be sung, the authorities decreed, they should be about Chairman Mao and his glorious revolution.

That was just the start of Mawen's suffering. Her husband died not long after their songs did. She moved to another village to be with her son, but soon after she arrived he fell from a boat and drowned. Mawen continued to live with her son's family for a short time, but when her son's widow decided to remarry, she was cast away.

I was crestfallen when Mawen told me her story of love, loss, and abandonment. Having treated a number of patients for a condition called Takotsubo cardiomyopathy, a sudden weakening of the heart muscle that is often called broken heart syndrome, I knew what emotional stress can do to a person's health.

"After such tremendous heartache," I asked her, "how did you find the strength to go on?"

"How does the river go on?" she asked.

"Well, I suppose it just does," I said. "That's what rivers do."

"That's what I did as well," she replied. "My second son came to help me and I simply got through it. I had lost one part of my family, but another part was together, so I should not be sad."

I walked away on that day thinking that Mawen was the strongest woman in the world, and feeling as though there was little I could do to emulate her emotional fortitude. What I have since learned from Mawen and others in Longevity Village, and what has been exceptionally well reflected in international medical literature, is that a positive mind-set isn't just about *feeling* happy. The way we *think* about our lives is perhaps the biggest factor in how our bodies will respond to the conditions of our lives.

Indeed, if we are going to change our lives for the better, we must first change our *minds* for the better.

SMILES AND LAUGHTER. That is what I came to expect more than anything else when I would visit Mawen in her home in Bapan. Already 107 years old when we first met, she was feisty, funny, and ferociously dedicated to helping others understand that life, while not always easy or fair, is too wonderful a thing to waste with sorrow.

I cannot recall a single moment in our time together that she was not smiling. When our families would sit together for a meal, her laughter carried over the muddled drone of our multilingual dinner conversations. Whenever I would pass her home, she would always wave to me, point to her mouth, and smile.

No one lives to be so old without experiencing some tragedy, and given the ups and downs of all of our lives, it's common for centenarians to have experienced devastating periods of misfortune. Yet researchers studying the personality characteristics of people whose lives reach one hundred years and beyond have found these individuals are almost always happy, optimistic, and resilient.

I've become convinced the mere *act* of smiling might help us live longer. One clever study showed that baseball players who smiled on their playing card photographs lived seven years longer, on average, than those who looked stern. Happiness has also been shown to be a key factor in marital stability, which in turn has been shown to be a key factor in long-term health. Another study demonstrated that a person's propensity to smile in childhood and yearbook photographs seemed to correlate with their likelihood of staying married.

I knew Mawen wouldn't have any photos of herself from when she was a teenager, but after hearing about the connection between getting "caught smiling" and long-term health, I was eager to ask her about that period in her life. Did she remember it as a happy time?

It was still quite early in the morning when I arrived for that visit. Mawen was already up, eating soup at her family's low-slung dining table in a crisp blue shirt and sparkling rhinestone headband. As usual, she was smiling ear-to-ear.

"Doctor!" she exclaimed as I walked into the room, and before I could offer so much as a greeting, she began to list the various ways in which she was, as she put it, "in perfect health." "I wake up every morning with energy," she beamed. "I am completely pain-free. My toes and fingers all work as they should! I can't walk as far as I once did, because I do get tired, but I can still get around."

"And you're still smiling!" I observed.

"Always," she said.

"Even when you were very young?"

"Always!" she repeated.

"Even when things were very hard?"

"Always," she said once more. "And those are the times in which smiling is most important, don't you agree?"

I very much agree. The conscientious decision to flex the muscles of your face into a smile releases neuropeptides (which work to fight off stress) along with dopamine, endorphins, and serotonin ("feel good" transmitters which are also released when we eat tasty food or engage in sex).

Just seeing Mawen smile always made me smile, no matter how I might have been feeling at that moment. We have that effect on one another. Smiling is contagious, after all.

We can even have that effect on ourselves. One of the foremost advocates for mindfulness in the modern world, the poet and peace activist Thich Nhat Hanh, has spoken extensively about the power of a smile. "Sometimes your joy is the source of your smile," he has observed, "but sometimes your smile can be the source of your joy."

Almost everyone I know has some sort of morning routine. And almost everyone's routine includes standing in front of a mirror for at least a few moments to brush their teeth, comb their hair, or look over their outfit. Next time you're standing in front of a mirror, smile at yourself. Make it as sincere a smile as you can muster. Then make that a habit.

That's how habits start: one simple action. When we feel success, we're far more likely to repeat the action leading to that success, and we're also more likely to try to take another small action that might, eventually, lead to another healthy habit. If you make a habit of starting each day with a smile for yourself, I can assure you, the difference you will feel will be incredible.

THERE IS AN old parable, variously attributed to several Asian cultures, about two monks, a master and his apprentice, who come upon a river. When they arrive, they find a rich woman who demands they carry her across the water. The older monk takes the woman on his back and, even as she moans and complains, he ensures not a single drop of water touches her. For miles thereafter on their journey, the younger monk stews and sighs. Finally he cannot take it any longer. "Master," he asks, "how could you let that woman treat you in that way?" The older monk replies: "Why should I care now? I stopped carrying her miles ago. Why are *you* still carrying her?"

Mawen had always adhered to the philosophy espoused in this story. She was certainly not free of the sorts of pain, suffering, trials, and tribulations that mark so many of our lives. But when it came time to deal with these misfortunes, she let go of her previous hardships and was more able to deal with whatever came her way.

I wondered for a long time how that could be. After all, when Mawen lost her family, she literally had nothing at all. She returned to Bapan in her mid-seventies without a single Chinese yuan and had to begin rebuilding her life from the ground up. In those days there was nothing like social security in China. The economy was in shambles.

But Mawen told me she simply had a choice to make, and it was a simple one: Survive or perish. And when you consider that our evolutionary compulsion is to survive, it makes it simpler still. The way Mawen saw it, her needs came down to the basics: food and water, shelter and security, love and belonging. Once she'd returned to the village and taken up residence with another child, she had everything she needed. Content in that way, she was able to go on.

None of this should imply that Mawen wasn't interested in striving for more, or that we shouldn't do the same. Mawen most certainly did continue to work to better her family's fortunes. In those years, she accomplished a great deal, but she didn't *need* to do that in order to put a smile on her face. It was already there.

That's why today, when she speaks of her husband, her son and even the daughter-in-law who ultimately cast her out of her home, Mawen likes to focus on the times in which her life experiences *exceeded* her expectations. And because her expectations were always quite simple, it's not very hard for her to embrace those memories.

When we strip away the inessential elements of our lives and focus on what really matters most, it's a lot easier to feel content. And while it's certainly not easy to switch our perspective from "I want more" to "I have what I need," it can have a drastic impact on our health.

BUDDHISM TEACHES THAT the root of all unhappiness is desire. This does not mean that we should desire *nothing*, but rather that we should focus our desires on that which is most important to us—family, health, and safety—and leave all feelings of entitlement by the water's edge. In this way, whatever life brings us beyond our expectations becomes a source of joy.

That's what makes Mawen's lessons important, not just for finding and maintaining happiness, which is vital to good health, but for creating other positive changes that are conducive to healthier living. When we're burdened by trying to *maintain* the inessential things we have (let alone procure more) we have less time, space, and energy to devote to making the real changes that are so vital to creating healthier and happier lives.

I've been there. I'm still not sure all of the factors that led me to become such a workaholic but I'm certain a big part of it was that my "provider and protector" instinct seemed to go into hyperdrive after the birth of our first child, and only increased when each new child arrived. I wanted to create an economic condition in which my growing family could survive any catastrophe. But the better we did financially, the more I was conjuring greater and greater potential catastrophes in my mind.

The only solution seemed to be even more work. Instead of taking time off to enjoy my children and support my amazing wife, I was back at the hospital performing surgeries within a day of their births. I didn't give myself time to relish those first precious days of their lives.

That was the status quo. Like so many other people I know, I'd bought into a backward sort of logic that told me that if my family was my priority, work had to take precedence over everything else. But as much as I believed my work was worthwhile, I could not in a week at the hospital derive the same joy and fulfillment as I achieved in a single hour with my children.

And, of course, I wasn't at all in the same sort of perilous economic situation as so many other people in this world. One of the greatest benefits of traveling to places like Bama County is the privilege of being reminded of my privilege. I can't seem to return to my home without feeling immensely blessed by everything I have, and all the opportunities that have been afforded to me simply for having been born where I was, when I was, into the family I had, and in the community in which I was raised.

At the same time, as I've spent more and more of my time outside the relative opulence of life in the United States, I've felt an increasing heaviness as I've recognized that none of the trappings of a modern, Western, "comfortable" American life have actually contributed to greater happiness. Somewhere along the way, many people in our society began associating happiness with "having more." The truth is these ideas aren't related at all. And while not mutually exclusive, they are quite often at cross purposes. That's how it was in my case.

Just because we can have more doesn't mean we need more. And this doesn't just go for money or physical *stuff*. It also applies to how we make our day-to-day choices. Just because I have the opportunity to attend many social events doesn't mean I need to. Just because I have a choice of thirty different soups at the store doesn't mean I need all this variety. Just because I have access to more medical care doesn't mean I need it. The fact is that, in many cases, the condition of having more choices makes us substantially *less* happy.

All of this is why, shortly after my first visit to Bapan and continuing today, I've embarked on an extensive effort to simplify my life, and I have invited my patients along on this journey. This process is always ongoing, and it is as simple as continually asking ourselves two basic questions:

First, do we have what we need? Next, do we need what we want?

These questions are applicable to all parts of our lives. They apply to our physical needs (food, clothing, and shelter); our emotional desires (love, happiness, and friendship); our time (for sleeping, working, eating, and socializing); and even our digital lives (e-mail, social media, subscription services, and digital "property" such as audio and video files).

When we ask these questions, we can better separate our needs from our desires, and that allows us to begin the process of decluttering our lives, which is an absolutely vital step toward stress relief. It also empowers us to say "no" to things we have or want, but do not need.

That is not to say that all wants are bad. Much to the contrary. The problem is that, all too often, our lives are too cluttered by things we do not need to fulfill any new desires (or, even worse, to address new needs).

Through this process, my patients and I have discovered that we can bring more meaningful experiences, much-needed time, and even new possessions into our lives without any added stress. These days I limit my daily to-do list to three things above my basic work-life schedule. Today, for example, I reviewed a scientific manuscript for publication, wrote a few pages for this book, and scheduled an upcoming medical lecture. Because of this, I still had the time, space, and attention for my son Jacob when he asked me to help him with some homework. Instead of fretting about letting him down, because I was too busy, I was able to give him the time he needed, and which I *want* to be able to give him as his father.

This isn't a habit reserved for those with economic privilege. My patients include police officers, copper miners, and single mothers who are maintaining multiple jobs to ensure their children have what they need. And, of course, the people who inspired this practice live in a poor, rural village in China and work harder than anyone I've ever met.

Although their opportunities are limited, the villagers all know that by staying out in the fields even longer, or by taking on a loan for a piece of modern farm equipment, they *could* have even more. But the happiest, healthiest among them have chosen to limit their expectations and desires, leaving more room for the things in life they really want and need.

The large sign, festooned with the images of all of the living centenarians and celebrating the town's status as Longevity Village that we saw as we walked into Bapan, helps tell an important lesson about the difference in mind-set that exists here.

In Bapan, youth is appreciated, but age is *exalted*.

Da Yi, a government official who oversees the Jia Zhuang Township, which includes Bapan and ten other villages, told me that when he was growing up, "we didn't know that other places didn't have the same number of centenarians we did. We thought that's just how it was everywhere. But even then we knew that our centenarians were special. To live to be that old a person must be very healthy and wise, and so they are like a hero."

The idea of centenarians as local heroes was fascinating to me. When I was growing up, I was certainly taught to respect my elders. But when I thought of a hero, it was almost always a portrait of youth, like a professional ballplayer, an Olympic distance runner, or an astronaut.

Rather than thinking of older people as heroes, I was impacted by the prevailing Western view that suggested the elderly are often fragile, slow of mind, and hardened of spirit. Perhaps even more influentially I was bombarded, as most of us have been, by a culture that worships youth and a youthful version of beauty.

But no matter how you look on the outside, it's how everything is working on the inside that really counts, because regardless of wrinkles, gray hair, or any other cosmetic feature of your outside body, there's a tremendous difference between calendar age and *biological age*.

When I first met Da Yi, for example, I was convinced that the local "mayor" of that part of Bama County was in his mid-thirties. True, he dressed in sharp, trendy clothes and had a full head of thick black hair, but if he had struggled to walk up the village's steep steps or groaned when he lifted himself from a low chair, I would have likely reconsidered my guess at his age. Because he walked about the town without ever seeming to slow down and seemed perpetually energetic and eager for another adventure from early in the morning until late at night, I never questioned my assumption.

One day, though, while we were waiting for someone in our party to purchase some fruit for lunch, Da Yi asked about the age of my youngest daughter. She was seven at the time, and I told him so.

"Ah!" he replied. "My youngest is almost that age."

"Daughter or son?" I asked.

"Not daughter or son," he said. "Grandson!"

When Da Yi told me he was fifty-six, I could hardly believe it.

I thought I had learned my lesson, but a few days later, when I met a woman named Yan Hong, I made the exact same mistake.

Yan Hong, who lived with Longevity Village's newest centenarian, 101-year-old Matao, had salt and pepper hair and some light wrinkling around her eyes. I pegged her to be in her mid-to-late fifties.

"So how has your grandmother's health been lately?" I asked her.

"Not grandmother," Yan Hong laughed. "She is my mother. I am eighty years old!"

Yan Hong could see my surprise. Beaming with pride, she told me she still worked in her family's farm plot every day and had no plans of stopping any time soon. "Maybe in twenty years," she said. "But I've never had any sickness, I don't feel tired, and I like to work, so I'm not sure why I would ever stop."

It occurred to me, as I watched Yan Hong skip down the stairs off to collect some food in her garden, that she'd probably never been told that there was a certain way people in their eighties were expected to feel and act. That's because none of the centenarians in Longevity Village think about "growing older," let alone worry about it. There is no fear in the passing of another year.

We should follow their example. Although our age is often a tremendously large part of our mind-set about "where we're at" in life, it is a perfectly pointless measure of who we are.

Inherently we all do know this. Not everyone "acts" his or her age after all. Almost all of us have met children who are "old souls" and adults who act like kids. The number of times we've circled the sun has far less to do with our well-being than we typically think.

To take advantage of what science *actually* tells us about aging, though, we have to get into the habit of ignoring the other messages we get about what is "supposed to happen" as the calendar flips from one year to another. To this end, it's important to know that a lot of the social and cultural messages we get about age, sometimes from our own friends and family, are wrongful and can even be harmful. And these messages can even come from those we trust most to provide us accurate information about our health and well-being.

Just prior to giving the keynote lecture at a large medical conference organized by Northwestern University in Chicago on the subject of "How to Thrive to 100," a surprising number of conference attendees I met said something to the effect of, "who in their right mind would want to live to be one hundred?" That sort of response completely ignores the fact that a person who arrives at one hundred in good health and happiness most likely enjoyed great health and happiness when they were sixty, seventy, eighty, and ninety, too. And these were other doctors!

But even doctors are products of their environments. They have been bombarded with messages that make getting old look awful and spent their careers examining people who have confirmed through their struggles the idea that getting old is a painful, lonely, confusing, and frightening ordeal. Of course, healthy people don't spend a lot of time at the doctor's office, so what doctors don't see nearly so often are the examples of healthy aging that have inspired me to believe that calendar age is little more than a condition of our mind-set.

Instead of simply denouncing those who respond with malice toward the idea of getting old, I try to change their minds. I tell them about Fauja Singh, who took up marathon running when he was 80 and continued running distance races well past his hundredth birthday. I tell them about Teiichi Igarashi, a former lumberjack from Japan who summited Mount Fuji for twelve years in a row, starting when he was eighty-nine. I tell them about Georgina Harwood, who celebrated her hundredth birthday by jumping out of an airplane and then swimming with sharks off the coast of South Africa. And, of course, I tell them about the amazing people of Longevity Village who stand as proof that the process of growing older isn't something that should be feared, but rather something that should be craved.

All of the centenarians I've spoken to in Bapan have told me they are living the *best* years of their lives. Those in the village who were not yet one hundred longed to get there. And looking forward to golden years that have the potential to be truly golden might be one of the best things you can do for your health *right now*, no matter what age you are.

There are tremendous health advantages to anticipation. Researchers have demonstrated, for instance, that the mere knowledge that a vacation is coming makes people feel happier at work and that the mere expectation that sleep is on its way lowers blood pressure. It's hardly a stretch, then, to conclude the mere belief that growing older is a positive thing might be a significant influencer of good health.

In a study of 660 older Americans, researchers found that regardless of socioeconomic status or even age, people who embraced the aging process and felt like life would continue to get better lived nearly 8 years longer than those with a more pessimistic view about aging and the future. That could be because pessimism can actually deteriorate our DNA—or, to be specific, the curled ends of nucleotides, known as telomeres, which cap our chromatids. Telomeres are sort of like helmets for our chromosomes, and when they deteriorate in a process known as shortening, it leaves us vulnerable to aging-related diseases.

Telomeres are perhaps the best evidence that the number of candles on our birthday cakes doesn't have to correlate with the way we feel and act. Sure, we're all getting older according to the calendar, but research has shown we can slow and even stop telomere deterioration by dealing in healthy ways with the stresses of our life.

With all that in mind, doesn't it simply make more sense to enjoy the idea of adding another number to our age, rather than allowing it to be a distress? The way in which we choose to perceive and deal with stress is, after all, a tremendous marker of biologic age. Studies show that those who embrace stress actually live 17 percent longer. In contrast, as measured by telomere length, it appears that people who don't effectively manage high levels of stress age their bodies by nine to seventeen years.

Mawen didn't need to know anything about telomeres to understand that unhealthy stress doesn't help her maintain her health, nor does it help change things she has no control over. A few years ago, as part of the government's efforts to bring better, safer homes to the people of Bapan, a new row of buildings was erected right in front of Mawen's home, blocking out her beautiful view of the river and mountains beyond it.

"Didn't that make you angry?" I asked her.

"It doesn't matter," she said. "I know that the river is still there. And now I have another excuse to take a walk to see it."

WE HAVE SO much going for us in the Western world. And yet our very definition of happiness is often caught up in having and doing *even more*. The challenge is that, in most cases, all that *more* we're searching for costs *more* money.

So we work. Not to meet our expenses now and in retirement, necessarily, but often in hopes of a life of "just a little bit more." And that can come at a tremendous cost to our short- and long-term health, and thus our actual happiness, because most of us don't really like our jobs.

Only about 13 percent of Americans say they enjoy going to work. And if you consider how *much* work we do, that's really a shame. The average American works more than 47 hours per week, and 1 in 5 of us work more than 60 hours each week. Even if you're just working a 40-hour week, though, you're spending about a third of your waking hours at work.

Back in 1988, researchers in Israel conducted interviews with more than 800 working adults who had recently undergone a physical examination, asking detailed questions about their work situations. Twenty years later, when they checked back in with the study's participants, the researchers found that those who reported having little or no social support from coworkers were nearly two and a half times more likely to have died during the study. That's a tremendously important finding that puts unsupportive work environments on par with smoking in terms of long-term mortality.

That's one of the many reasons why I often ask my patients about how they feel about their working lives. When I recently did so with a patient named Mary, she told me it felt like it was getting harder each morning to get up and go to work.

"Why do you do it then?" I asked.

That wasn't a question I posed lightly. During my health crisis, I worked insanely long hours. In retrospect, I really wish someone would have asked me this same question.

Mary looked at me incredulously. "Well Doctor," she laughed, "if you'd like to pay my bills for me . . ."

Mary was a high school principal. That's one of the most stressful jobs I can think of. Formerly she'd been a science teacher.

"Did you struggle to pay your bills when you were a teacher?" I asked.

"It was certainly harder back then," she said. "But I made it work."

"Did you like teaching?" I asked.

A huge smile washed over her face. "I really loved it," she said. "And I miss it quite a bit."

Mary cut me off before I could ask the obvious follow-up question. Returning to teaching would be very difficult, she said, because she had "grown into" her administrator's income. She and her husband had purchased a larger home and had spent several weeks each summer over the past few years traveling in Europe.

"Do those things make you happy?" I asked.

Yes, she said. She loved her new house and cherished her vacations.

"OK," I said, "is the happiness those things bring to you greater than the happiness you derived from teaching?"

Mary paused for a moment, shrugged, and offered up a sad smile. "Not even close," she said.

I glanced down at her charts. Her blood pressure was worryingly high.

"We're going to talk more about your blood pressure," I said, pulling a pad of paper out of my white coat pocket, "but first I want to offer you a prescription."

In my very best doctor's handwriting, I scribbled the words: "Consider going back to teaching."

I'm not going to downplay the obstacles Mary and her husband would face if they decided to relinquish the very sizable difference between her salary as an administrator and what she was making as a teacher. These are very complicated decisions, and there aren't often clearly right or patently wrong answers.

I do know that, for me, balancing my surgery schedule with more research, teaching, and writing has been vitally important to my overall well-being. It is considerably less lucrative, but it brings much more meaning and happiness to my life.

A colleague recently observed, though, that the decrease in my surgery schedule didn't seem an even trade for all the other things I was doing. "I really don't get it," she said. "It's clear to me that you're less stressed, but it actually seems like you're doing a lot more than you were before."

Indeed, I told her, I probably was. But given that it was a better balance of things I enjoy, I no longer felt stressed. I'd actually found

I could do more, be happier, and have more energy left over to dedicate to other parts of my life.

At least 70 percent of all visits to the doctor are for stress-related ailments, and in my clinic it is probably the same. Unhealthy stress also plays a major factor in the lives of the patients I see suffering from high blood pressure and heart conditions.

My friend, Dr. Rachel Lampert at Yale University, has found that feelings of sadness, anger, stress, impatience, or anxiousness increase a person's risk of suffering from atrial fibrillation nearly 600 percent. Worse, these effects aren't temporary; the risk can be carried over to the next day.

I've seen this effect in the lives of my patients, with tragic results. One of them, Luis, was just fifty-two when he began being treated for an irregular heart rhythm. A few months into his treatment, he was laid off at work.

"At first I didn't feel anything," he told me. "I just went home and took my dog for a walk and thought, 'Well, I guess I'll take a few days off and then go look for a new job.'"

The next morning, though, Luis said he woke up feeling incredibly anxious, as though if he didn't get started looking for a job right away he might miss the one opportunity out there for a man of his age and skills. He jumped into the shower, shaved his beard, threw on his nicest suit and tie and headed out the door, all before the sun had even risen. He was sitting in his car, still in the driveway, trying to figure out where he was even going, when he felt a sharp pain in his left arm. Fortunately, he was able to call 911 on his mobile phone and was able to give the emergency dispatcher his address before he passed out. Emergency responders were able to shock his heart back into normal rhythm. Before he was discharged from the hospital, though, we agreed that he would engage in some simple stress relief exercises any time he found himself feeling anxious about his work situation. After all, I reminded him, a health collapse was certainly not going to help his chances of finding work.

The reason for these sorts of adverse effects starts in the hypothalamus of our brains, which links the nervous system to the endocrine system and is evolutionarily programmed to respond to changes in our environment that might threaten our well-being. When this happens, the hypothalamus sets off a chemical chain reaction that sends adrenaline and cortisol coursing through our bodies. This raises our heart rate and blood pressure. And while that can be a healthy response now and then, it can be exceptionally unhealthy, and even deadly, if it happens too often or goes on too long. That's not just a problem for heart health, either. Unrelenting stress can cause migraines, aggravate diabetes, contribute to skin conditions, and increase depression and anxiety.

What I've found among the people of Bapan is an exceptionally low level of perceived stress. Indeed, they live their lives in a way that minimizes any unnecessary emotional strain; often unaware that the things they do, day in and day out, are helping them in this regard.

One day in Bapan, for instance, I found myself picking vegetables with a man named Li Yu, who told me he was fifty years old.

"Do you ever get tired of working in your field?" I asked.

Li Yu stood up, wiped the sweat from his forehead with the back of his hand and scratched at his chin.

"*Nà méi bànfǎ a!?*" he asked. "What choice do I have?"

I told him that I had heard of many people traveling to find work in cities like Guangdong, the coastal province surrounding Hong Kong and Macau.

"It's the same," he said after a bit of thought. "Either way I would be doing what I have to do to survive, but here in the village I get to be out in the sun, breathing the mountain air, and close to my family."

"It seems like very hard work," I said.

"It is hard work," he said. "By the time I am back at my home, though, I don't think about how hard it is. I am always feeling satisfied about what I have accomplished during the day."

Li Yu's field was on the opposite side of the water from his home in the village, past where the river splits, temporarily, to go around a small island. At this place, the river is shallower, and on most days of the year the water was low enough there that he could walk across without having to go upstream to the village footbridge.

Because he had to step slowly and cautiously to avoid falling and being swept away by the water, it probably took Li Yu longer to walk across the river than to use the bridge. But, he said, he enjoyed it more.

"I let the water cool my feet and wash away the dirt from the field," he told me. "I always feel better when I walk across the river."

Li Yu probably didn't realize that there is a scientific and psychological basis for the way he felt after his daily trek, but there is. It was once widely believed that once a brain circuit is established for one purpose (let's say, for instance, the rote mechanics of hand washing) the circuit could be used only for that specific purpose. Today an emerging field of research called neural reuse is showing these circuits can be used for multiple purposes. One of the most fascinating lines of inquiry in this field of science is the question of how a single circuit might be responsible for controlling both physical actions and metaphorical connections. In a series of experiments designed to better understand this effect, a team of researchers from Michigan showed that the mere act of hand washing can help people feel better about decisions they've made, rid themselves of emotions like anger, and lessen their judgments of the actions of others. Both literally and figuratively, you can wash your hands of work. I have no doubt that Li Yu was experiencing this same effect when he would walk across the river, letting the passing water wash away the difficulties of his workday.

We all have stressors in our lives. The key to living well with stress lies in how we perceive and manage it. For instance, it might help to do things that physically and technologically separate your body from the source of your stress. I have a patient who has found that parking his car a few blocks from work means that, at the end of the day, he has no choice but to take a walk, a well-established way to shed stress and anxiety and a great physical metaphor for getting away from the office.

For some of my patients, the answer has been to change the way they commute to work. Driving is one of the biggest stressors in many people's lives; a trip instead by train, bus, ride share, or taxi can be an incredibly productive way to reduce our stress. For some of my patients the answer has been to go on a social media "fast." (It's amazing how much unhealthy stress Facebook and Twitter can add to our lives.) Others still have resolved to only check their e-mail at certain times of the day. (Researchers have demonstrated that taking regular breaks from the constant deluge of e-mail messages can significantly lower stress.)

Disconnection isn't just important for those of us who are tethered to work through technology. The people of Longevity Village have long recognized the importance of leaving the farming on the farm. Even though field work takes up the vast majority of their days, conversation over lunch and dinner rarely trends in that direction; instead it most often centers on news from around the village, like who is having a new baby or who is moving up among the ranks of local mahjong players. And in the sweltering summertime, people gather in the few twilight hours after dinner to walk together in the cooler twilight air, swim along the shallow banks, or sing together in one of the pagodas that overlook the river.

"There is no good reason to talk about work during these times," one villager gently chided when I asked about his farm during a pause between evening songs in a colorfully painted pagoda upstream of the village. "There is time enough for that when we are working. This is a time for us to play."

MANY OF US spend at least some of our lives engaged in exercise and athletics, but most of us don't *play*. We might be moving and, indeed, we might even be having fun, but we're usually not creating, imagining, and improvising. And these, I've come to learn, are vital attributes for a healthy mind and body.

This certainly isn't a lesson I expected to learn in China, whose citizens are more popularly known for their rigorous attention to rules than their playful attitudes toward life. But as I listened to Mawen describe the ad-libbed songs she and her husband would sing as they worked in the fields, it occurred to me: Play was central to their entire day.

Slowly, I began to recognize many other things that the villagers in Bapan did to be playful throughout their days. Lifting pebbles from the field and hurling them toward a faraway tree like a golfer aiming for a flag on a practice range. Telling stories that have been told a thousand times before and listening as though this was the very first time. Laughing and joking at one another's foibles, like when someone stumbles and their basket of veggies is overturned, without malice or resentment.

When we treat work as play, we change the very nature of work. We rob it of its power to stress us and deplete us of our energy. We turn it into something we enjoy, look forward to, and covet. That is good for us because play is *so* good for us.

There is a lot of natural evidence that play is an indicator of good health. Play among wild bears, for instance, occurs when they are well fed and not under pressure from hunters. Robert Fagen, who wrote a book on play, has observed that when bears and other animals are playing, they are also thriving and living longer, too.

One of the best ways to transform stress is to make sources of potential unhealthy stress a source of play.

Does your commute to work drive you nuts? Try listening to the books or podcasts you never seem to have time for while driving. Do office rivalries or politics really annoy you? Try organizing an office softball team, putting together an annual March Madness NCAA basketball tournament pool, or organize an office service project at a local homeless shelter. Truly: Be playful.

For me, being the cardiologist on call for my hospital is very stressful. Between surgeries I am often running from one end of the hospital to the other consulting on patients or running back and forth to the emergency room. To make it a game, I try to see how many steps I can log on my iPhone during these on-call days. While it may not seem like much, having my phone count the steps I take crisscrossing the hospital all day long goes a long way toward defusing the stress of being on call.

It's also crucial that the things we do to play are, indeed, play. If you participate in an indoor soccer league every week, but leave each game feeling angry at yourself, your teammates, your opponents, or the referee, then you're turning something that is supposed to be healthy into something that is quite unhealthy. Likewise, if you only go to the gym begrudgingly because you don't enjoy working out, then you're absolutely setting yourself up for failure. If the way you're working out isn't fun for you, you're almost guaranteed to quit at the first setback. At its best, exercise is something we should look forward to. Positive anticipation, after all, is a vital part of a healthy mind-set.

THE REASON IT took me so long to address my eating struggles was that I'd always assumed my exercise habits were canceling out my poor diet. Sure, I'd acknowledge to myself, this brownie ice cream sundae evening snack is giving me a megadose of sugar, carbs, and fat at a time when my body is least capable of handling it, but I'm going to wake up tomorrow morning and run for five miles, so it's all even.

One of the most important things I learned in the village is that, when it comes to our health and happiness, there's really no such thing as an even trade. We can't just cancel one bad action with another good one. Unfortunately, though, we've been programmed to believe that we can. Even worse, it's likely that when we take positive actions to improve our health (like running) we're more likely to forgive ourselves for other actions that are detrimental (like evening ice cream sundaes).

A lot of the things we do to try to counter the impact of bad decisions can actually cause even *bigger* problems for our health and well-being. We see this often around the New Year, when people focused on undoing what they've done to their bodies during a voracious season of holiday eating suddenly take up an aggressive exercise regimen, often running. It turns out, though, that about a quarter of all new runners have injured themselves within their first 23 miles. Half of those are still so injured after 10 weeks that they're not yet ready to engage in a moderate running workout. And about 5 percent have to get surgical treatment. That's a tremendously bad consequence of good intentions.

A lot of medications that we've come to associate with improving health outcomes can have similarly negative unintended consequences. Take statins, one of the most frequently prescribed drugs for lowering cholesterol. When researchers from UCLA looked at data from tens of thousands of Americans who participated in the National Health and Nutrition Examination Survey between 1999 and 2010, they discovered that those who take statins eat 10 percent more calories, 14 percent more unhealthy fats, and gain an extra 10 pounds in the process. This might not be a disconcerting finding if increased appetites were among the chemically induced side effects of these drugs, but statins aren't known to make people hungrier or crave junk food. Rather, it appears that once statins have been prescribed, patients often feel as though the drug will take care of their problems for them, and thus spend less time concentrating on eating healthy.

The totality of all of this is that we have to change the mind-set that tells us we can counteract negative actions with positive ones. As I tell my patients, "You can't exercise your way out of a bad diet." That's almost always going to be a zero-sum or even negative-sum game.

Thankfully, there's a bit of a hack for this, because not all behaviors that are fun are also unhealthy. What kinds of things do you like doing that are also either healthy or have little impact on your health one way or another? Whenever you have the impulse to "reward" yourself with something that isn't healthy, like junk food, pick a healthier and ultimately more rewarding thing instead. For me, that's definitely spending time doing something athletic with my family. That completely satiates my impulse for a reward and, of course, has the added bonus of creating even more meaningful time with the most important people in my life.

Will you slip up? Yes, sometimes you will. I certainly have. Let bad days become good data. Learn from your mistakes, do everything you can to understand what went wrong, and make a plan to prevent the same thing from happening again.

WHEN A FRIEND of mine announced he was changing jobs a while back, his boss asked him what kind of cake he'd like at his going-away party. Mindful that many of his colleagues were trying to develop healthier eating habits, my friend asked instead for a veggie tray.

Five years later, he still doesn't hear the end of it. "Any time I get together with an old coworker it seems they've got to bring up the veggie tray," he recently told me. "It's always been sort of sweet, like 'it was bad enough that you left us, but then you robbed us of cake, too,' but honestly, if I'd known that was what people were going to remember about the party, I probably just would have let them eat cake."

In Bapan, no one would think twice about inviting a neighbor over to commune over some fresh vegetables or fruit. They've got a deeply ingrained connection to food, after all, since it's more than likely that the food in their homes was harvested by their very own hands. If they happen to harvest too much, the work they expended is wasted, so it's not unusual for someone to invite a friend over to make the most out of some extra carrots, apples, or peppers before they begin to spoil.

Why is inviting someone over for alcohol, caffeine, or sugar-packed pastries socially accepted while inviting them over for veggies might be considered weird? Why do we celebrate birthdays with cake instead of a complete healthy meal? Why do we bring cheese-laden casseroles to potlucks instead of a bowl of roasted pumpkin seeds?

Little by little, we can take actions that fly in the face of conventional wisdom, especially if what we do conventionally isn't particularly healthy.

That's not just limited to food. One of the most promising workplace trends I've seen over the past few years is the advent of the standing desk. Sitting for long periods of time, after all, has been linked to colon cancer, muscle degeneration, poor leg circulation, and higher mortality rates. And while standing might not be the health panacea that some people seem to think, it is most certainly a move in the right direction. When you're on your feet already, after all, it's tremendously easier to add more movement to your day than when you're slumped into a chair.

Despite this, woe be the person who is the first in her office to push back her chair, prop up her computer on a stack of books, and, well, *take a stand*. In almost every situation I've heard of, that person has had to endure a sizable dose of mockery before, one by one, others catch on to the tangible benefits of getting out of their chairs.

If you're going to take actions that are better for your life and the lives of people you care about, you're probably going to have to endure a bit of ribbing, too. And if that's going to happen the best thing you can do is to simply adopt the mind-set of not really caring what those people say.

If you were to get on a video conference call with me right now, chances are good that you'd see my head and shoulders bobbing up and down the entire time. That's because when I'm in the office but not seeing patients, I'm often strolling along on my treadmill desk. I know that some people think it's funny and strange, but I long ago stopped caring about that. Instead, I've decided to put myself in a mind-set that says that anything I can do that is healthful for me and not hurtful to others is worth doing, and even worth some teasing I might endure as a result.

WHEN YOU STOP caring what people think about you, it's a lot harder to get angry. That's a really good thing, because when it comes to our health anger is pretty much the worst emotion in the world.

Some of the most compelling evidence of that fact can be found on Twitter, of all places. If you were to "retweet" every post you read that expressed hate or hostility, you'd be sharing more than just negativity; you'd actually be helping to pinpoint the places in the United States where heart disease is at its worst. In fact, when researchers crunched the numbers they found they could use angry Tweets to predict regional heart disease rates with uncanny accuracy—better, in fact, than they could by looking at other factors like smoking, diabetes, and obesity.

While the idea that we might be able to predict heart disease through social media messages is relatively new, the understanding that anger correlates to poor health has been well established. In study after study, going back for decades, anger has been shown to increase the incidence of serious heart problems, worsen insomnia, aggravate digestive problems, prompt headaches, and worsen depression.

When I told my dear friend, Bapan village historian Fu Ji, about these studies, he shrugged. Anger between villagers, he said, isn't something that he's had to worry about for a long time.

"Here, anger is rare," he said. "There is a warm relationship between everyone. There is no mistrust. There are no suspicions. Everybody trusts everybody."

I knew that my friend had suffered grievously during the Cultural Revolution, a period of political upheaval between 1966 and 1976 that included heinous attacks on intellectuals and the building of the cult personality around Mao Zedong that still exists today.

"What about during the Cultural Revolution?" I asked.

"That was a very bad time," Fu Ji said. "Everybody was fighting against each other and yes, then there was a lot of mistrust. People had no idea what they were doing."

At the time, Fu Ji told me, there were forty-two people from this tiny village who were teachers, including him. Prior to being known as Longevity Village, he noted, Bapan was called Education Village. "This made us an even bigger target for persecution," he said.

"Are there people who were aggressors against you who are still in the village?" I asked.

Fu Ji sighed. "Interestingly, the people who participated in attacks and persecuted others have all died now," he said. "I think it was the justice of heaven."

"But not right away," I responded. "These people lived among you for a long time, didn't they?"

"Of course," Fu Ji said.

"But weren't you angry with them?" I asked.

"No," he said. "We did not yell at them. We did not seek retribution. In our hearts we knew the crimes that they had done, and so we were not close to them, but there was no anger. What happened at that time was terrible. There was no reason for us to relive it."

Fu Ji's story isn't unique. In all of the time I've ever spent in Bapan, I've never seen anyone angry. There is irritation. There is frustration. There is sorrow. But the emotion I'd described as anger, meaning that someone is perceptibly provoked and clearly agitated, doesn't seem to be part of village life.

I thought for a time that this was just part of putting on a good face in front of visitors. But when I spoke to the families of village centenarians, they told me that they couldn't remember a time, not once in some very long lifetimes, in which they'd seen their elders act angry. And everyone else in the village, they noted, took their emotional cues from the elders.

Sure, they knew what anger was, they'd seen it on television and in the reactions of tourists from other parts of China whose experiences were not matching their expectations. But as an emotion, it was as foreign to them as another language.

I KNEW IT was unlikely that I could completely clear all negative emotions from my life, but I wanted to reach a place where it occupied the smallest possible space in my reservoir of emotional responses. I explained this to Mawen and asked for her advice.

"Zěnme?" I asked. *"How?"*

"Shēnhūxī, fàngsōng," Mawen responded. "Breathe deeply and relax."

At first I understood this response literally. There has been lots of research on how breathing impacts our emotions. In fact, I can probably make you feel a certain emotion, just by instructing you to breathe a certain way.

Want to try that? Breathe and exhale slowly and deeply through your nose, with regular intervals between each breath. Relax your rib cage. Repeat for 30 seconds or so.

Did you feel . . . happy?

Most people who are not already in a heightened emotional state who go through this exercise will say that just a half-minute of deep breathing increases their sense of being happy to some perceptible degree (and sometimes to a very significant degree). That's what a team of psychologists from Canada and Belgium reported in a study published back in 2002. First, the researchers asked a group of people from all walks of life to remember times in their lives in which they felt sad, happy, angry, or afraid, then recorded how their breathing patterns changed. They noticed almost immediately that people tend to have very similar breathing patterns in response to these emotions.

We can use what we know about breathing to bring our emotional state into balance, especially when it comes to managing our anger. My patients and I have recognized together, however, that while these techniques help dissipate feelings of anger in rather short order, they don't necessarily *prevent* feelings of anger. After all, anger *is* a natural response to situations in which we are physically threatened; the physiological state it invokes (including what evolutionary psychologists call "anger face") is perfect for communicating to the affronting party that they've done something wrong and is good for getting our bodies ready to, say, fight off a saber-toothed predator.

The problem is that our brains also use anger as an emotional response to social situations in which there is no real danger and in which the affronting party can't even see our "anger face." And when not needed for survival, the physiological conditions brought on by anger can be unhealthy and even dangerous.

Consider, for instance, how you feel when another driver honks and makes angry gestures at you. What good would anger do you in that situation? At best, it's a relatively benign response that subsides as the offending driver disappears into traffic. At worst, it incites you to respond in unwise or dangerous ways. But neither response is actually *beneficial*.

Breathing in response to a feeling of anger triggered by this sort of situation isn't a bad idea, but it would certainly be better if our emotional response were not anger to begin with. A healthier response might be amusement at the colorful cast of characters in our world, sadness that this person feels the need to act out, or curiosity as to what might be going on in this person's life to elicit that kind of rage.

Does that sound like an impossible state of heightened emotional consciousness? I thought so, at first, too. That's why it initially occurred to me that "you must breathe" might not be the best advice for preventing anger. But when I considered the source of the advice was someone who couldn't recall ever feeling anger and who wasn't remembered by others to have ever expressed anger, I recognized that breathing wasn't necessarily a response to anger, but a protective measure against it.

Mindful breathing, long before we get angry, is a great way to ensure that we don't give in to an unhealthy response when ultimately provoked. That's what mindfulness master Thich Nhat Hanh advises. His mantra, "breathing in, I calm my body; breathing out, I smile," isn't a response to stress, but rather a proactive preventative measure.

An important thing to remember is that none of these ideas about managing our anger are emotional repression, which has been shown to create greater risk of cardiovascular disease and even cancer. Rather, the goal we should strive for when it comes to improving our current and long-term health is to be better connected to stronger, more powerful, and more beneficial emotions. And when we exercise those emotions, they're more likely to materialize when we need them. And all it takes to get started is one conscious breath.

That's something *any* of us can do—all by ourselves. But if we truly want to live longer, happier, and healthier lives, it's important to recognize that we can't do it alone.

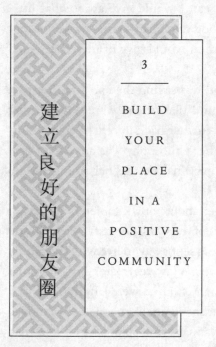

3

BUILD

YOUR

PLACE

IN A

POSITIVE

COMMUNITY

"The people around me desire my well-being. If they did not, I would not permit them to be around me."

—MASONGMOU

MASONGMOU HAD NOTHING, THEN EVERYTHING, THEN NOTHing again.

Yet when I came to sit beside her on the back deck of her home, looking out over the peaceful Panyang River, she told me she was content because she had *always* had everything in the world she could possibly want or need.

At the time we first met, she was 108 and confined to a wheelchair as a result of a fall on a slippery tile floor a few months before. Up to that point, Masongmou had been fully ambulatory. But when I asked her whether she was sad that she could no longer completely take care of herself, she raised a steady, slender finger and corrected me.

"Here," she said, "we all take care of each other."

"My life," she continued, looking out over the river, "is a direct result of how I have always helped other people. If anyone in the village did not have enough to eat, I would give away what little rice, vegetables, fruit, or other food that I had so that those without enough food could eat. Each time I would give of myself it would give me more life—a longer life."

At first I thought Masongmou was saying she had done her part for others, and so she didn't mind that others were now helping care for her. But as she reached further back into her personal history, I came to understand that her philosophy was not one of give and take, but one in which the cycle has no real beginning and no real end.

Born in the village of Poyue, about four kilometers upstream from Bapan, in the waning years of the Qing Dynasty, Masongmou was twelve years old when she was sent away to live with the family of her future husband, who was ten. A few years later, Masongmou's in-laws both passed away, and their children were left destitute.

When Masongmou and her husband finally married, they moved to Bapan where they had been granted a small plot of land. They had nothing, not even enough food to eat while they worked on their first harvest. When the people in the village saw how desperate they were, they gave them their extra pots, clothing, and food. When word spread of their plight, volunteers arrived from across the region to build them a small hut.

Masongmou's husband was a gifted fisherman and their harvests were bountiful. "We saved everything that we could and worked very hard so that we could buy more land," she said. "With all of our land, at harvest we were able to generate more food than our family needed."

Eventually, they became the richest family in the village. But Masongmou said she never forgot the help they'd gotten from other villagers when they first arrived. Whenever someone else in the village was in need, she understood it was her responsibility to help them. She had everything she could have dreamed of, and sought to share her good fortune with others.

Everything changed when Mao Zedong took power. In 1950, less than a year after the establishment of the People's Republic of China, Mao presided over an uncompromising agrarian reform law that redistributed land to peasants who did not have any. (Ultimately, though, the land would all become the property of the state, as it has remained ever since.)

Party members openly encouraged the killing of landowners, landlords, and those who were considered "rich peasants" like Masongmou and her husband. Rather than have police and party officials preside over the executions, though, the party encouraged peasants to take matters into their own hands. At least one million landowning individuals and their family members were killed. Some estimates put the number of dead at greater than 4.5 million.

Masongmou and her family were spared.

"Everything was taken from us and we would have starved to death," she said. "But all of the people we had previously helped secretly came to our rescue and made sure we had enough food to eat, even though this might have brought condemnation to them as well. It was only through their generosity that we were able to stay alive during these years."

Masongmou's story confirmed for me something that health researchers long ignored, even though (or perhaps because) it is so patently obvious: Healthy communities make for healthy people.

If you want to live a longer, healthier, and happier life, it's every bit as important to pay attention to your community as to pay attention to what you eat. In fact, it now seems clear that our communities are an even *more* important factor in our current and long-term health than what we eat.

"Community" means different things to different people, but at its most basic it is the people with whom you surround yourself. That can be your family. It can be your friends. It can be your neighbors. It can be people with similar interests with whom you maintain a connection. And, especially in the digital age, it doesn't necessarily have to be geographically located; it can be people you feel connected to, via social media, even if you have never met each other in person.

However you choose to define it, though, one thing is clear: When we strengthen our community, we strengthen ourselves.

BY THE TIME I met Masongmou, she and her husband had long since rebuilt their lives once again, and had lived together in happiness and contentment even as they both became very old. He died at the age of ninety-five after being hit by a car in a nearby village. She had then gone to live with their youngest son.

There is a tradition in this part of China in which children build a coffin for their parents when they reach the age of sixty. That might seem morbid, but I've come to learn that it is not in any way because the children believe their parents are close to death. Much to the contrary, the practice is intended to let aging parents know that everything has been taken care of, and they do not have to worry that their deaths might be a burden upon their kin. When I asked Masongmou about this, she laughed.

"My son has already prepared three coffins for me," she said. "The first two became old, molded, and fell apart."

That, I learned, was quite common. The wooden coffins tended to last ten or twenty years before they began to rot and fall apart.

Masongmou wasn't eager to die, but when the time came, she said, she would be ready and content. And the way she saw it, death would give her a special opportunity to give back to her community once again.

As is the case throughout most of China, each home in the village has a shrine. Ancestor worship—*jìzǔ*, in Mandarin—is deeply ingrained in Chinese culture and rooted in Confucian philosophy. Many Chinese believe the deceased have a special ability to do favors for the living, and can even shower their living descendants with blessings and fortune. That's one of the reasons they make offerings to their ancestors, even burning paper money to make sure departed loved ones have plenty of cash to spend in the spirit world.

There's a powerful social effect to this belief system. For in addition to whatever heartfelt familial obligations children feel they have to their elders, the constant presence of an ancestral shrine in people's homes is a persistent reminder that the way parents are treated in this life won't be forgotten when they pass on. This certainly isn't to say that the Chinese wouldn't be devoted to their parents anyway; the Chinese take elder care quite seriously. Indeed, China is the only nation in the world I know of where children can be prosecuted for not visiting their parents often enough. But the spiritual coupling of familial responsibility and self-interest turns out to be a powerful motivator that helps keep the elderly engaged in the lives of their families in ways we sometimes fail at here in the United States.

What results from this belief is an understanding of *mutual need* that lasts until a person's very last breath. Nearly every home in Longevity Village is inhabited by three or more generations from the same family. Each generation relies on others for emotional, educational, social, and economic support. Children are expected to give grandparents and great-grandparents love and companionship. Elders, in turn, are expected to help children learn to be good scholars and good people, and to help out with cooking and cleaning as much as they are able. *Everyone* is important in a meaningful way. That's what community is all about.

WHILE WALKING ALONG the village's upper road one day, I saw an older woman shoveling concrete into a wheelbarrow then pushing it to where a young man was lifting it, bucket by bucket, to the top of a building with a mechanical pulley. The home belonged to Mawen, and I guessed the woman must have been Mawen's daughter-in-law, which would almost certainly put her in her seventies or eighties.

When I stopped to help her, she thanked me for my time but expressed concern that I might hurt myself.

"I had the same concern about you," I said.

"But I have been doing this all my life," she said. "You are a doctor, and I'm worried that you have never done work like this before."

When I asked her how long she would be moving the concrete for, she told me the job would be over soon enough.

"That's my grandson up there," she said, gesturing to a man who was perched on the roof. "I told him that if he doesn't hurry up on his end, I'm going to come up there and he'll have to come down here, and he likes it much better up there!"

It would be very hard to overstate the health benefits of feeling capable and connected to others in this way. Regardless of age, people who feel as though their communities don't need them are at increased risk of depression and high blood pressure, and have been shown to be more susceptible to all sorts of other ailments. Disconnection is the beginning of loneliness, and that is even worse for our long-term health: Research suggests people who feel lonely are more than twice as likely to suffer from Alzheimer's disease.

The indelible feeling of capability and connection I saw in the people of Longevity Village isn't always present in the lives of my patients. And that's a real problem, because loneliness and social isolation can actually be more dangerous to your health than smoking or even being an alcoholic, and perhaps twice as dangerous as being obese.

Yet even as our nation spends hundreds of millions of dollars on antismoking campaigns, billions treating alcoholism, and hundreds of billions battling the health consequences of obesity, we haven't treated *connectedness* as the public health priority it should be.

Perhaps it's no wonder that between 1985 and 2004—years that saw an explosion of cable television channels, a proliferation of angry talk radio shows and, of course, the emersion of the Internet from relative academic obscurity to the predominant media force of our time—the number of Americans who don't have a *single* person with which to discuss matters that are important in their lives nearly tripled.

The trappings of our modern society can leave us helpless, frustrated, isolated, and sick. But these parts of our lives can also be tools we use to strengthen our connections with friends and families, and build our communities in ways that will help us feel stronger, more important, and healthier than we've ever been.

Masongmou is the perfect example of this. When I next returned to the village, her wheelchair was nowhere to be seen. "My family and friends told me that I'd been in it long enough," she explained. "They said, 'You need to start walking again' and so I did."

RECENTLY, WHILE SEEING a patient who was suffering from a heart arrhythmia, I inquired about who she considered to be part of her support network.

"Oh, Rosemary," she said with a sly smile.

"Is that your daughter?" I asked.

"No, my cat," she laughed. "The best friend a woman could have!"

We both got a good chuckle out of this, but I did want to know that she also had some *human* support. Did she have regular visitors? Did she ever feel lonely?

"How could I feel lonely?" she asked. "I have three wonderful kids and seven beautiful grandchildren. Someone is over at my place almost every weekend."

"And the weekdays?"

"Oh, well," she said. "Those days *can* get pretty lonely. But I wouldn't want to complain."

I assured her that it was perfectly reasonable to feel lonely after four or five days spent with her cat as her only companion. We discussed the health implications of loneliness, and some strategies for addressing it.

As it turns out, a lot of people think that there's a difference between being lonely and being isolated. But at least in terms of health impacts, that's not the case. That was a key finding of a 2010 study of social relationships and mortality by Brigham Young University's Julianne Holt-Lunstad, whose research has helped paint our growing disconnection to one another as the health epidemic it truly is. What Holt-Lunstad has found is that the increased likelihood of death for people who reported *feeling* lonely was statistically similar to that of people who were socially isolated, no matter how they professed to feel about that isolation.

I've definitely noticed in my patients a tendency to downplay or deny feelings of loneliness. The deeper we get in conversation, though, the more it's clear that this is a problem that impacts a lot of their lives, sometimes without them ever realizing it. And that's increasingly problematic in a world in which isolation is increasingly the status quo.

Elders are far from the only people who get lonely. Loneliness is actually one of the most common emotions among Americans of every age, so much so that many people don't even realize that they are, in fact, chronically lonely. That feeling isn't something to be ashamed of; it's simply our body's way of telling us that we need to work a bit harder toward the goal of connecting with our community.

How do we do that? I often tell my patients that the best way to cure their loneliness is to cure someone else's. After all, when two isolated people are together, they're no longer isolated.

On occasion I've asked patients who appear to suffer from loneliness to do a bit of "homework," using the Internet to research recreational sports teams, interest groups, political organizations, book clubs, volunteer organizations, and lifelong learning classes that they might be interested in joining. All of these avenues are great ways to meet new people, and the mere act of looking into these opportunities is often enough to get people to try them out.

When Mika, an Iraq war veteran in his mid-thirties, came to me with high blood pressure and an abnormal heart rhythm, I quickly began to suspect that social isolation might have something to do with his health problems. Since leaving the Army, Mika had worked hard to stay in shape, and when he noticed it was getting harder to beat back his beer belly, he resolved to stop spending so much time in bars. The problem was that barhopping with friends had been a big part of his social life. When I asked him if he'd be willing to look into some local groups that share his interests, he agreed, and by the next time I saw him, he'd joined a skeet-shooting club. While we also worked together on other aspects of his well-being, I'm convinced that his regular involvement in the club has been a major factor in his improved cardiovascular health without any medications or procedures.

IT WAS A year or so after my first arrival in Bapan that I began to notice that no one seemed to share hurtful gossip about other people in the town. I was coming to know many of the town's residents very well, but I hadn't been let in on any of their dirty laundry. It was refreshing to come to know a community in this way, and very different from what I'd come to expect in other communities, even those I might describe as "tightly knit."

When I shared this revelation with Masongmou, though, she seemed confused.

"Why would someone spread negative information about another person?" she asked.

"In my experience," I said, "when someone is hurt or upset with another person they tend to tell others about it."

"But that makes no sense," she said. "What is appropriate is that they tell the person with whom they are upset."

If only it were *always* so, I thought. In the United States, it seems, if we encounter someone we feel we can't really trust, it is easier for us to tell others rather than address it directly with the person.

Holt-Lunstad has spent a lot of time studying this phenomenon, and what she's found is potentially devastating for our health and well-being.

"About half of our associates," she told me, "fall into the 'frenemy' category."

Social scientists, Holt-Lunstad said, call these folks "ambivalent friends," but the definition is about the same. These are people in whom we find some value, so we tend to want to maintain the contact, "but we're never sure we can count on them or trust them," she said.

These aren't just casual acquaintances, either. Holt-Lunstad and others have found that our immediate family members are just as likely, if not *more* likely, to act like ambivalent friends. After all, it's a lot harder and more heartbreaking to try to cut an irresponsible sibling out of our emotional lives than it is to drop a person with whom we have no familial relationship.

While it might seem logical that there's a mental health consequence to surrounding ourselves with people we don't really believe we can trust, (and indeed there is), Holt-Lunstad and her colleagues have found there is a very real and very disconcerting consequence to our physical health as well.

First, they asked more than 100 people to wear blood pressure monitors over the course of 3 days, and noticed right away that when these volunteers interacted with their "frenemies" their blood pressure went up, even if the specific contact with that person was positive in nature. In fact, interactions with "frenemies" raised the volunteers' blood pressure even more than interaction with people that they didn't like whatsoever.

Short of the people I've come to know in Longevity Village, I can't think of anyone who doesn't have at least a few ambivalent friends. What's worse, in speaking to my patients and others about this issue, it seems that most of us feel there are people in our lives who don't lift us up. If you're surrounded by people who don't lift you up, won't treat you right, and don't appreciate your goals, you stand a much lower chance of living a long, healthy, and happy life. Open up space in your life by limiting your exposure to negative people, and bringing in positive people.

That's what my patient Jared did. "In the past," he told me, "I would meet someone new and think, 'Wow, I'd love to get to know that person better,' but I never acted on that impulse because I simply didn't have the room in my life. Now that I've limited how much time I devote to the people who are more negative forces in my life, I've found I have a lot more opportunities to share my life with people who are really positive forces on me."

SOMETIMES JUST THE mere act of changing one's life for the better is enough to repel negative influences. When we decide to engage in the world with positivity, it seems that positive people appear all around us.

The truth is, though, that even the people who treat us with the love, respect, and encouragement we deserve won't always understand the changes happening in our lives. If you've ever seen the hurtful expression on someone's face when you decline to eat something they've made but that isn't healthy, you know how this can go.

So how do we make positive life changes without negatively impacting the communities we've built around us? Well, perhaps the best way is to remember one of Aesop's most famous fables, a story that long ago made its way from ancient Greece to eastern Asia. In China, they call it *wūguī hé tùzǐ*, or "The Tortoise and the Hare."

Now, it's true that slow and steady does *not* always win the race, but when it comes to building and maintaining supportive communities around us, it's best to think in terms of distance rather than speed. When we make small, positive, and consistent changes to improve our health and well-being over time, it's a lot less likely to seem confusing, inconsistent, or hurtful to those around us. That makes us a lot more likely to succeed.

Why not just sprint out of the gate? Well, for one thing, we tend to leave the people around us behind, and those people can be vital to our success. Going through a weight-loss program with a group of friends or family members, after all, increases the chances of long-term results by 275 percent. That's a tremendously good reason to start slow and steady, and thus improve the opportunity to run in a pack.

Of course, we can't force people to join us in our quests to improve our well-being. And preaching to our friends and family members about how to live longer, healthier, and happier lives isn't going to help them, or us, actually achieve that goal. The biggest and most lasting changes are always rooted in community and positivity.

When a social psychologist named Paschal Sheeran looked at 144 studies involving 11,712 people trying to change unhealthy behaviors, he found that the most effective approaches used positive self-affirmation, where people focused on important values, attributes, or social relations when making changes. Even with good intentions, telling the people we love they need to start eating better isn't likely to make them eat better. If the way we talk to them causes them to feel guilt or shame, it might make them eat more. And that definitely isn't doing any favors to our relationships, which means it's not doing us any good as we strive to meet our own goals.

To be clear, I've never discouraged my patients from making changes in their lives aimed at greater health and well-being, even if those changes are big and sudden. But I do suggest that when they do make a change, they remain conscientious about how it will impact the people around them who are already acting as positive forces in their lives.

That's the village way.

IN THE MORNING when she would rise, Masongmou would help her son prepare a simple breakfast of corn porridge and vegetables. At midday, when he would break from his work, they would eat together again, generally some rice, hemp seed soup, legumes, vegetables, and tea. And at the end of the day, they'd eat together once more, perhaps having a small bit of fish, rice, and pumpkin greens, and often joined by neighbors or other family members.

Historically, almost everyone in Bapan eats with their family during every meal. And as I watched family after family engage in family meals, I realized that it had been a long time since my whole family had sat together for three meals in a single day.

According to the USDA's Economic Research Service, about half of all US food dollars are spent eating fast food and other meals outside of the home; that's twice as much as it was in 1970. And researchers estimate that half to two-thirds of all fast food is ordered at the drive-thru window, accounting for the fact that about one in five meals we eat are consumed in a car. Meanwhile, more than ever before, we're eating alone. More than half the time we eat, it's by ourselves. The only meal Americans eat more often with others than we do alone is dinner, and even then about a third of our suppers are consumed solo.

These aren't positive trends, and they're not trends limited to Americans, either. A few years back, researchers in Europe looked at 25,000 adults over the course of 20 years to try to better understand how people's lifestyles can impact chronic diseases like cancer and diabetes. They found that older adults who most frequently ate alone consumed a lot fewer vegetables than those who were living with someone, and were much less healthy as a result.

At my home, our lives have gotten no less busy, but now that we've shifted our mind-sets to recognize that a lot of that busyness is a choice, we've made another choice: As much as we can, we cook and eat together. We set the same time for family dinner as often as possible and assign each child one night per week to plan and help prepare the meal. It's still not perfect, but we spend a lot more family meals together now than we ever did before, and we eat a lot better as a result.

There are lots of people, though, who don't have the option of having family meals like we do, not just because of busy schedules, but because they are part of the growing segment of our society that lives alone. Among their ranks was a young resident at my hospital named Julie, whose parents, siblings, and childhood friends all lived thousands of miles away.

"I still have family dinners almost every night," Julie beamed when we were discussing the subject of how to help our patients build stronger and more supportive communities.

"But I thought your family was in New York," I said.

"My parents are, and my sister is in Miami," she said. "But we get together for a meal every Wednesday night over video. We've been doing it for years."

"And what about the other nights?"

"When I'm not working? On Mondays a bunch of the single residents get together at someone's house. We trade off cooking for one another," she said. "On Tuesdays I eat with a woman I met at the senior center where I volunteer. Friday and Saturday are usually date nights; that's not really family dinner, but it's nice to get out and eat with someone new."

I tried to envision Julie's calendar in my head. "So that leaves Thursdays and Sundays, right? What do you do then?"

"I've gotten pretty good at inviting myself over to people's houses," she said. "By the way, what are you and your family doing this Sunday?"

I loved Julie's pluck. That's really what it takes to wrestle ourselves away from societal norms that aren't healthy. If you're not quite ready to invite yourself over to someone's home for dinner, though, there are plenty of other less audacious ways to make sure you're always eating with someone. One great way to do this is to take turns with a coworker preparing lunches for two. Meals are also a great time for collaboration with colleagues, although I strongly advise my patients not to schedule high-stress meetings during mealtimes. Even simply taking a meal you prepare for yourself onto your balcony, front porch, or another space that will bring you into contact with your community can be a positive step toward feeling more connected with others. And if you can afford it and are feeling up to it, whenever you're about to make a meal for one, make it for two instead, and then go knock on the door of a neighbor's home and invite them to share it with you.

The benefit isn't just social. When the people with whom we are eating understand our health goals, they can become an important part of our motivation and accountability structure. It's easy, after all, to discard rules we've set for our food choices when no one knows. When our friends and family members are in the loop, even if they don't share the same practices, they can help us stick to our plans.

OUR COMMUNITIES BEGIN with our partners, and when it comes to making healthy changes in our lives, there's no one more important.

A patient of mine named Louise learned this the hard way. In her late forties, Louise noticed that she was starting to develop chest pains, and that she would sometimes get winded just walking from her downtown parking lot to her office a few blocks away. That led to a cardiac workup which ultimately resulted in several heart stents to open up blocked arteries.

It was the proverbial wake-up call. Louise's son was just about to graduate from elementary school. The idea of not being there for his high school graduation was too much to bear. In my mind, that was a very good thing: Louise had something to live for and look forward to that was still quite far down the road.

"The odds of seeing that dream become a reality are going to be a lot better if you change how you eat," I told her after her workup, which included a review of her diet. "Your life literally depends on your daily food choices."

Louise was convinced. And she began in the most logical place: the grocery store. When Louise would go shopping for herself and her family, she would work hard to make conscientious choices about what she was putting in her cart. Soon, though, her husband,

Gary, started to notice that a lot of the things that had been staples in their cupboards, like Oreo cookies and Little Debbie snack cakes, were disappearing.

When Gary confronted her about this, she told him that she had decided they were going to eat healthier. His response was to get in the car and drive directly to the store, where he loaded up a hand-basket with all the things Louise hadn't purchased. "If you're going to eat healthier," he told her that evening, "I'm very happy to support you, but these foods make me happy and I'm going to keep eating them."

What Gary didn't understand was that in bringing those foods into their home, he wasn't supporting his wife at all. Rather, he was significantly increasing her risk of another stent, heart attack, or even a young death. Without a home free of junk food, her chances of eating healthy were really low, just like how having easily available liquor in the home of an alcoholic would drastically reduce that person's chances of staying sober.

What Louise didn't realize, on the other hand, is how much she had hurt her own chances of success because she didn't invite the most important person in her community to become involved in her effort to get healthy; she'd simply made a decision for both of them.

They remained angry at one another for weeks, and Louise found it hard during this period to focus on making healthy choices. She found herself sneaking cookies, and when Gary noticed she was doing this, he used it as an opportunity to score points in their ongoing argument. That just made Louise angrier, which in turn made it even harder for her to focus on eating healthy.

"I know it seems incredibly dysfunctional," Louise told me. "And the thing is that it was such a strange place for us to be. Over twenty years of marriage we'd always been so compatible. Up to that point, I don't think we'd ever had an argument that lasted a whole day, let alone several weeks. It was awful."

I'm happy to report that Louise and Gary have worked out an arrangement that is helping them both work toward a common goal. When Gary is out of cookies and cakes, he has agreed to go to the store to get them for himself, and this actually has reduced his consumption of junk food, since it takes extra effort on his part to get it. He has also agreed to keep these sorts of food in his basement office, where Louise is unlikely to just happen upon it when looking for healthier foods.

NOT EVERYONE CAN arrive at this sort of détente. That's one of the key reasons why so many diets fail. About 84 percent of people who begin eating a vegan or vegetarian diet, for instance, eventually opt to consume at least some meat. A lot of people point to this statistic to suggest that perhaps we're not *meant* to eat that way. On the other hand, there are millions of people who do eat this way, very successfully and very happily.

What's the difference between those who succeed and those who don't? In large measure, it is support. One of the key findings of a study of 11,000 Americans' eating habits was that those who switch to vegetarianism or veganism need support to avoid feeling as though they are "standing out from the crowd."

Changing lifelong eating habits takes a level of willpower and determination on par with what is expended by people training for marathons and triathlons, or who engage in ultra-athletic training such as CrossFit. Going it alone is almost impossible. If we're going to be different, as it turns out, we really need people to be different with.

The healthiest communities are those in which we're surrounded by people with mutual goals and values, and who act accordingly and consistently. Here I'll use an example from my own faith: As a member of the Church of Jesus Christ of Latter-day Saints, I'm encouraged by my church's doctrine to eschew tobacco, alcohol, and any other potentially addictive substance. As tobacco and excessive alcohol account for approximately a half million deaths each year in the United States, this is no doubt a key reason why studies have shown that Mormons may live up to 10 years longer than others.

Because I have these guidelines and am surrounded by friends, family members, and other fellow church members who share a dedication to them, I've never had any trouble avoiding these things. Truly, it has never even crossed my mind to drink a beer, smoke a cigarette, or take an unnecessary drug.

But without guidelines and people around us who share a dedication to them, Mormons are as prone as anyone else to make bad decisions about what they consume. You can see this reflected at pretty much any modern Latter-day Saints social gathering. There you'll often find lots of sugar, plenty of processed grains, and endless soda pop. Perhaps if we mindfully approached processed foods such as these, as well, we'd live not just 10 but up to 20 years longer! Better still, if Mormons could set an example of healthy eating, then perhaps people would come to think of us in *that* way, rather than as the people who ride around on bicycles with white shirts, dark ties, and black name tags.

One of the things that people often notice about Mormon missionaries is that they tend to go around in pairs. In fact, with the exception of when they shower and use the restroom, they're never supposed to be alone during their missionary service. Some people might find this strange, and maybe it is, but given the purpose of the mission, it makes quite a bit of sense. Missionaries are often young adults, usually just out of high school or in their first years of college. They are asked to adhere to a very strict regimen of proselytizing, study, prayer, and community service. Could they succeed alone? Maybe some, but certainly not most.

Partnership works. And if you're engaging in a life-altering experience, having a partner by your side can be an invaluable way to ensure success. This is especially true if you partner with another person who has strengths where you have weaknesses, and vice versa.

That's what my friend LuAnn did. The successful baker was quite active at work each day, but she knew in order to really turn the corner on her well-being, she needed to get in some more *intensive* exercise each day.

"I tried to do it on my own," she told me, "but it's just not my thing. I'm not an exercise person."

Her friend was, though. And as it turned out, that friend felt like she could really use someone to talk to on a regular basis.

The solution was pretty easy; they walked and talked. "She knew that every day I was counting on her to help me keep walking when I'd get tired and want to take a break," LuAnn said, "and I knew that she was counting on me for the conversation, which was something I really enjoyed, too."

The arrangement was so beneficial that when LuAnn's friend moved away, the pair continued to exercise and chat together every morning over the phone while wearing headsets.

When we surround ourselves with people who are dedicated to the same healthful practices as we are, we're *so* much better off. Yet, as Louise's story shows us, sometimes we can't even find that sort of protective insulation within our own homes.

Whenever her life was in flux, Masongmou once told me, she would go door to door to connect with her friends and neighbors to explain the challenges she was facing and the solutions she had decided upon. "If I didn't tell people, then these decisions would have been secrets," she said. "And no one could have helped me if I was keeping a secret."

That's why, when my patients make lifestyle decisions in an attempt to better their health, I advise them to approach the most important people in their life and do four things:

1. Describe the change they're making.
2. Explain why they are making it.
3. Assure their loved ones that they don't expect them to make the same change unless they want to.
4. Request their support and understanding.

The "DEAR" approach has worked wonders for many of my patients who are seeking to build a supportive community as they make healthy changes to their lives.

DURING HER VERY long life, Masongmou has been a wife, a mother, a farmer, a singer, a political outcast, and a local celebrity.

What she has never been is confused about her importance to her community. In her direst times, she knew she could count on them to care for her. In her best times, she knew they needed her.

There's almost no way to quantify the benefits to her health, happiness, and longevity. Instead, I can only tell you what she told me one day as we sat on her back deck and watched the lazy currents of the Panyang River trickle by.

"When your life is important for others and their lives are important for you," she said, "then you are very rich."

And with that, a woman who had been confined to a wheelchair when we first met got up and headed for the kitchen to help her son.

"You never stop moving!" I called after her.

"And I won't!" she called back.

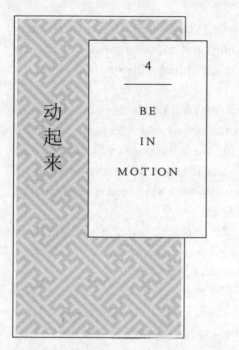

4

BE

IN

MOTION

动起来

"What I have done all my life is not exercise. It is simply my life."
—MAKANG

MAKANG WAS AGHAST WHEN THE VILLAGE MATCHMAKER INtroduced her to her future husband, a boy from a poor farming family in Bapan. She knew the match had damned her to a life of toil in the sun and dirt.

"I was only ten years old and it would be many years before we were to be married," she told me, "but even then I knew his family was very poor and our life would be very difficult."

Like any child, Makang said, she had other kinds of dreams. "But my grandfather told me, 'This is a good arrangement because you will live close to our family.' By the time we were married when I was nineteen, I had accepted that he was right."

During those first years of marriage, Makang and her husband were desperately poor. "We never stopped working," she said. "I know now, though, that this is why I am alive today. All of that physical labor was very good for me, and I came to enjoy it more than anything."

Makang was 104 years old when I met her in the summer of 2012. She was small, wiry, and strong, and to me it seemed as though she was always on the verge of exploding with youthful joy, like a tightly packed firecracker.

During that first visit, we sat with our hands clasped and I looked into her enigmatic, energetic eyes as we spoke. Her home was humble but comfortable. She certainly was not poor by local standards. As one of the Longevity Village elders she had many guests from across China, and they often came with small red envelopes filled with gifts of money, called *hóngbāo*.

"It must be nice now to be able to relax," I said. "After all of those years of hard work, does it feel good to be able to sit here and appreciate what you have accomplished during your life?"

Makang sat forward, nearly lifting herself out of her chair. "Not at all!" she said. "I loved being in the fields. Until only last year I was in the fields every day, planting and harvesting food for me and my family, working every day for many hours. For me this was the very best thing in the world."

She stopped and looked out across the river. On the other side, farmers were hunched over in the dirt. "Every day I wish I could be back there, but my grandson has asked me to help our family at home," she said. "Here I still get to care for our family's pigs and ducks, though. That is good. I still prefer to be doing manual labor than sitting here meeting guests."

She paused again and looked back at me.

"I don't mean all guests," she said. "You are very nice." And we both laughed and laughed.

Whenever I'd see her after that, we would not sit for our chats. Instead I'd follow her around her house as she went about her daily chores, sweeping up, cutting vegetables, feeding the animals, or chasing around her great-grandchildren.

Makang never exercised a day in her life, at least not in the way that Westerners usually think about exercise. But throughout the years, like others in Longevity Village, she never stopped *moving*.

THERE IS LITTLE organized exercise in the village of Bapan. Children run footraces in the streets, set up hurdles with bricks and broom handles, and have swimming contests in the river. But there are no teams, uniforms, or coaches. No one criticizes form, or gives out ribbons to the winners.

Among the adults, there are no running groups. No gyms, either. And until quite recently, there wasn't even tai chi, the slow motion stretching and meditation exercises that are an emblematic morning ritual almost everywhere else in China.

Today, prodded by a national public service initiative to inspire more exercise among the people of China, there are growing factions of what I'd call "intentional exercisers" in the village. In the morning, small groups of women walk along the riverfront, hands swaying from side to side. One couple I know walks out to the island each morning with an old handheld radio for their daily aerobic workout. There's even a small group that, of all things, has taken up bullwhipping as an exercise; they gather together on a pavilion near the road overlooking the village and crack, crack, crack away for hours on end.

"No one did any of those sorts of things until quite recently," Makang once told me as we watched a young woman and her husband jog by in matching tracksuits.

The older residents of Longevity Village never intended to exercise, but they did *get* exercise. Lots of it. Indeed, almost every waking moment of their lives was spent in motion. In the morning, after rising to prepare food and eating breakfast, nearly everyone in the village would make their way to their family farming plots. For a long time, even those who had other jobs generally spent an hour or two in the morning tending to their crops. In rural China, land is owned collectively and provided through private use rights to individuals or families. If it is not used in accordance with the "public interest," it can be taken back and given to someone else, and many families zealously guarded their plots by actively farming their land, even if they didn't need to.

Throughout her life, Makang told me, she would be in her field every day from sunrise to sunset—strolling through the rows to spread seeds; bending to pick weeds or to harvest vegetables; toiling with a shovel or hoe. On the rare occasions when she would finish her day's work before the sun went down, she would still linger in her field, wandering among the crops to take stock of what would need to be done in coming days.

"I never felt tired," she said. "If I felt an ache doing one thing, I would just change for a while to do something else."

I once asked if she ever took a day off. "During the Chinese New Year, yes," she said. "And when I was young when my children were born I would take a few days to rest before returning to the field."

"But other than that?" I asked.

"No," she said. "Why would I?"

When she would harvest, Makang said, she would load her vegetables into a large basket, as heavy as she could possibly carry, and heft them onto her back for the walk back to the village, often with a few other people at her side doing the same thing.

"This was the thing that took the most energy," she said. "But it wasn't a long walk. Maybe we would sing a little song as we went to take our minds off of how heavy the vegetables were. When I would put it down my entire body would feel lighter again. I would have energy to keep working in the field again."

I've come to believe that a life of constant but not overly taxing movement, using every muscle group in our bodies, is a model for the very best kind of exercise there is. This should include short bursts of more strenuous activities. Fresh air. Friendship. And, most of all, enjoyment.

You don't have to toil in the fields of rural China to enjoy the benefits of exercising this way. You can have the benefits of this sort of exercise no matter who you are, where you live, or what you do for a living.

After all, our bodies were meant for movement. Not sports. Not bicycling or Zumba or fitness video games. None of these things are bad for us, of course, but few of us do these things *all day*. We exercise and then we stop. We play a sport and then we stop. We bicycle and then we stop.

It's really time for us to stop stopping. Because that's not what our bodies were designed to do. Our bodies are exquisitely fashioned for walking and running, hunting and gathering, climbing and crawling, swimming and swinging, and so much more. We are designed to be in movement all the time.

Studies show the average person now sits 9.3 hours a day. That is far longer than the amount of time we sleep. At least when we sleep, we're constantly in motion, arousing ever so slightly between periods of our deepest slumber to reposition ourselves before drifting back into our dreams. Even in our least conscious moments, our bodies are working to make sure we keep moving.

And when we're awake, our bodies are *screaming* at us to stay in motion. It's hard to rest in a solitary position over the course of a single hour without having some part of us fall into a state of pins and needles, which happens when pressure cuts off the blood supply to nerves throughout our body, rendering them unable to communicate with our brains. It only takes a few hours of immobility, a flight from New York to San Francisco, for instance, to significantly raise the risk of blood clotting, a condition known as deep vein thrombosis.

It can take as little as a single day of being sedentary for our muscles to atrophy; I've seen this in my post-op patients, and it's not pretty. It can take a week or longer to gain back the muscle strength lost over just one day in a hospital bed. It doesn't take much longer than a day of rest for our bones to lose density.

Yet even though our bodies are doing everything they can to tell us to keep moving, our modern world is giving us a set of very different messages. Sit don't stand. Drive don't walk. Let a machine do that for you. We have engineered activity out of our lives.

And, quite tragically, these messages have proven to be quite powerful. Over the past two generations, physical activity in the United States has fallen by a third. For every three hours in which your grandparents were in motion, you're likely only moving for two hours. Meanwhile, our livelihoods are demanding less from us in terms of physical activity. In the early 1960s about half of the jobs in private industry in the United States demanded moderate physical activity; today it's less than 20 percent.

In China, the drop in physical activity began later, but has been even more precipitous. Over the course of just one generation, the number of hours Chinese children spend engaged in physical motion has fallen by nearly half! These changes have been felt in Longevity Village, too. And this has certainly not escaped Makang's attention.

"I would like the next generation to have an even better life and also live a long life, but I cannot predict this," she said. "Things are very different now. When I was young all I did was work in the fields. Now so many people are going to the cities to work in offices."

Makang told me she felt stuck. She was happy for the changes that have offered more opportunities to her more than fifty children, grandchildren, great-grandchildren, and great-great-grandchildren. At the same time, she couldn't tell if those changes were making them happier or healthier.

"In time we will know," she said.

With that, she stood up. "I'm getting tired of sitting here," she said. "It's time to move."

ONE OF THE first things I noticed about the homes in Longevity Village was that there wasn't much in the way of furniture intended for sitting. The village centenarians often had a settee upon which they'd rest while meeting guests, and most homes had a few small chairs, the size you might expect to find in a kindergarten classroom, around a short dining table. That's about it.

I know of no home with a couch. Until recently, there wasn't anything resembling a park bench anywhere in the village. And when people come together for a wedding, or any other formal gathering, there weren't the rows and columns of chairs you'd expect at a Western ceremony; most people simply stood or rested in the squatting position common in much of Asia.

I used to sit all the time. In the morning I'd get out of bed and go right to my desk to check my e-mail. I'd sit for breakfast. I'd get in my car and sit down for the drive to work. I'd go to my office and sit while I worked on my computer. I'd sit during meetings. I'd sit while talking to patients. I'd even sit on a specially designed stool while I performed surgeries. Then I'd go home and sit on the sofa.

Today, I've become convinced that *sitting is the new smoking*. And by some calculations, in fact, it might be even worse.

By simply comparing the life expectancy of smokers with that of nonsmokers, then subdividing the difference by the average number of cigarettes a smoker will consume over a lifetime, researchers writing for the British Medical Journal noted that each cigarette reduces one's life span by an average of 11 minutes. The researchers' estimate struck a chord, and was repeated in articles and antismoking campaigns for years to come.

Far less attention has thus far been given to another estimate, also shared in a British Medical Journal publication but a decade later. Using similar methods, researchers computed the difference in life expectancy between heavy TV watchers and those who watch no TV at all, and concluded that every hour of TV watching after the age of 25 reduces one's life span by 22 minutes!

Yes, according to these estimates, both of which are admittedly crude, an hour of sitting down watching TV is as bad for you as two cigarettes! The way I see it, then, if you're going to sit down to watch a television show, it *really* should be a great program.

Of course, correlation is not causation, and untold other life choices that are shared among those who tend to watch a lot of television are certainly at play here. There could also be something about TV specifically that makes that activity worse than other forms of sedentary behavior, such as surfing the Internet, spending time on Facebook, or sitting at a desk at work. TV watching, after all, is often accompanied by sugary beverages and junk food snacks.

The overwhelming scientific evidence, though, tells us that *any* sort of prolonged sitting is simply bad for us. People who sit for long periods of time each day are far more likely to suffer from colon cancer, endometrial cancer, and even lung cancer. What's worse is that the research does not support the idea that exercise can negate the effects of lots of sitting. It doesn't.

I believe that, in just a few generations, we'll be looking back in disgust at how much time our current society spends sitting down. Yet even though there is little that can be done sitting that can't be done standing, most of us have yet to *take a stand*.

If you are a student right now, I'd love to see you take a stand by asking your teachers, school administrators, or student government to consider how to integrate more opportunities to stand up and move around in the classroom. If you work in an office, bring a milk carton to work and prop up your computer on it, then take a stand by encouraging others to do the same and talking to your employer about purchasing desks that are conducive to standing workers, and even better than that, for mobile workers. One place where many of us sit where standing could be beneficial, and even spiritually enlightening, is church; you can take a stand by talking to your religious leaders about whether "standing services" might be an appropriate way to worship. And once you're on your feet, you might find it's easier to dance, clap, and move about as part of the service, if that's an acceptable form of reverence in your place of worship.

One of the best places to start, of course, is in your own home. If you have a living room that is really more of a *sitting* room, consider making it a place where you're more actively engaged in the practice of *living*. Push back the couch. Add a treadmill, an exercise bike, a space for yoga, or even some free weights. Turn your floor into a putting green. Anything that you can do to get off your backside and onto your feet is a tremendous step forward.

THE CHINESE HAVE an old saying: *Fàn hòu bǎi bù zǒu, huó dào jiǔshíjiǔ*—Take one hundred steps after eating and live to ninety-nine.

This saying is well-known throughout China, but in much of the country, especially Westernized areas like Shanghai and Beijing, it has been relegated to little more than a quaint idiom. In Bapan, though, this rule is still treated with a significant degree of respect. After dinner, each night, the town is awash with energy as almost everyone leaves their homes for a stroll.

Big meals often leave a big surge of glucose in our blood. What happens with that glucose depends a lot on what we choose to do next. If we do nothing, our bodies will address these blood sugars with insulin, which feeds our fat cells and soon will leave us feeling hungry again.

But our muscles also need glucose to function, so if we simply go for a walk, the large muscles in our legs will quickly extract a portion of the sugar from our bloodstream, preventing insulin spikes and robbing our fat cells of the opportunity to gorge on the glucose. Instead of going to fat, the glucose is stored in our muscles as glycogen.

Essentially, we're filling our tanks. We can store about 90 minutes of activity-fueling glycogen in our muscles. That's about 2,000 calories. If our glycogen fuel tanks are already full, the excess glucose has to go somewhere else, but if we deplete those stores right after a meal then we can create extra space for that glucose to go.

That's what Dr. Loretta DiPietro of George Washington University found when she had a group of diabetic patients commit to taking a brisk walk after each meal. The post-meal walkers essentially prevented glucose spikes. And the walks didn't have to be long. In fact, three 15-minute walks after each meal were shown to be more effective than a single 45-minute walk each day. Indeed, frequent periods of low-impact exercise are almost universally better for us than longer periods of more sporadic exercise, even if it is more intense.

THE CHINESE GOVERNMENT has worked hard to convince its people to participate in regular exercise activities. The result in Bapan is that many people who are in motion all day long have come to believe that they probably need even more intentional exercise in their lives to be healthy.

By contrast, less than a third of Americans believe they are getting enough exercise. That's bad enough. What's worse is that most of those who think they're getting enough exercise actually aren't, and they're not even close. When using objective measures, rather than self-reporting, we find less than 5 percent of Americans actually get the exercise they need to live a healthy life.

Indeed, almost all of my patients initially overestimate the amount of exercise they are getting each week. As it happens, they also underestimate the amount of time they spend sitting each week and the number of calories they are consuming.

Recently, I asked a patient in her early thirties to tell me how much exercise she got each week. "I'd say a moderate amount," she said.

"That's a great start," I said. "What do you do?"

There was an awkward silence in the room and the woman blushed. "I honestly didn't mean to mislead you," she said. "I've always considered myself pretty athletic, but I guess I haven't really done much actual exercise since college."

I told the woman not to worry; she's definitely not alone. When I put pedometers on my patients, I generally see that they take about 3,000 steps per day, which is less than a third of what I'd consider sufficiently "active" to considerably impact one's health in a positive way.

In one of many studies in which this phenomenon has been observed, female participants overestimated the amount of time they were spending engaged in vigorous exercise by 500 percent, while men overestimated their vigorous exercise by nearly 700 percent! Meanwhile, men and women alike were sitting for more than two hours longer each day than they reported.

Part of the problem is clearly definitional. Many people simply don't understand what *vigorous* exercise is. When we're engaged in this sort of physical activity, our heart rates should significantly rise and our respiration should, too. More palpably, when engaged in vigorous exercise, most people are not able to say more than a few words at a time without taking a breath.

Another problem is our brains are simply less active when we are sedentary, especially when there is a screen in front of us. If you've ever lost yourself in a good movie, you know that it's easy to lose track of time when your brain activity is guided by what is happening in the film. If it can feel as though time stops when we're watching a show, it probably shouldn't be surprising to know that we have a hard time estimating how much time we actually spend engaged in any sort of screen activity.

Finally, the time we *schedule* for exercise tends to be a lot longer than the time we actually *spend* exercising. Whether we go to the gym, play basketball in a rec league, or meet with our friends for a regular round of golf each week, many of us spend nearly as much time preparing to exercise as we do exercising. When we do, we generally engage in exercises that only moderately raise our heart rate and respiration. After that we often get in a car and head home. By the time we've walked through our front door, our heart rate and respiration are normal again.

Whatever you're already doing to exercise is probably beneficial. So if you enjoy it, there's no reason you shouldn't keep doing it. But if you're looking for the kind of physiological effects that can revolutionize your life, a few trips to the gym each week probably aren't going to cut it.

Often when I tell patients this, they insist that they don't have time to work out *more* often. And what I tell them is they could probably stand to exercise a lot *less* often, and still get the same or even better results. That's because a lot of the benefits we derive from the exercise we typically do can be obtained from just a few minutes of vigorous activity each day.

For decades, researchers at the Cooper Institute, a nonprofit organization in Dallas dedicated to studying human health, have been compiling a database of health information and heath survey responses from tens of thousands of people who have had health checkups at an affiliated clinic. They've also been keeping an eye on public death records, looking for the names of former patients and, in particular, those who have died of heart disease.

What they learned doesn't at first seem that surprising: Patients who identified themselves as runners in the surveys were less susceptible to death than those who said they weren't runners. The risk of death by heart disease for runners was 45 percent lower and, as a group, runners lived about 3 years longer than people who did not run.

But here's the unexpected part: It didn't much matter whether they ran a lot or a little. Whether they were marathoners who put in dozens of miles each week or morning joggers who spent five or ten minutes each day taking a run around the block, the advantages in terms of longevity and reduced risk of heart disease were about the same. As it turns out, it's the combined benefit of a bit of strenuous exercise each day and continuous movement throughout the day that really impacts people's lives.

It's no coincidence that this is what Makang did. She spent nearly every waking moment of her life in her beloved fields, but only really felt as though she was exerting herself when she would lift a heavy bag of vegetables onto her back for the trip back to the village.

I have yet to meet a patient who cannot implement this philosophy of exercise and movement into his or her life. If you're already engaged in a job in which you're in motion throughout the day—as a chef, kindergarten teacher, or postal deliverer, for instance—then chances are that all you really need is a few minutes each day of strenuous exercise. A bit of high-intensity weight lifting, a morning run, or some living room aerobics would probably do the trick. The emphasis is that these things should indeed be strenuous, meaning you're working up a sweat and your heart rate is considerably up when you're doing it.

And it should be *every* day. If it's a morning jog, then it has to be every morning. If it's an evening bike ride, it has to be every evening. If it's a combination of things, then it has to be a combination of things you do week in and week out without really even thinking about *making* it part of your life, because it simply *is* part of your life. Monday martial arts. Tuesday tennis. Wednesday workout. You get the idea.

The easiest way to add this type of movement to your day is to make the world your gym. If you live or work in multistory buildings, for instance, all you need to do to get more strenuous exercise is to simply increase the use of the stairs. Unfortunately, a lot of people who try this relatively easy form of additional strenuous exercise end up giving up because they try to end their use of the elevator cold turkey. That's a recipe for failure. Instead, start by getting off one floor below your destination and add one floor at a time over a comfortable period of time. Then, once you've conquered the stairs, pick up the speed.

If you already engage in strenuous daily exercise, but are in a more sedentary job, then you'll need to find ways to keep moving at work. And this can be a bigger challenge for a lot of people, but it's almost always a surmountable one. The trick is finding the adaptation that works for you.

My choice of the treadmill desk allows me to work for hours on my computer and not even realize that I'm exercising. I have one at my home and one at work, and I set both to a very leisurely pace of 2.5 miles per hour. James Levine, a professor in the Endocrine Research Unit of the Mayo Clinic, has estimated that if sitting computer time were replaced by "walking and working" for just two or three hours a day, a person struggling with obesity could lose up to 65 pounds in a single year without making any other changes to his or her life.

Some people worry that walking while working would be difficult, but yearlong experiments have shown that workers who use treadmill desks are no less productive than their sitting counterparts, just healthier. Other studies have shown that people actually think *better* when they are moving.

But treadmills are certainly not the only way to go. Admittedly, they can be expensive and usually quite bulky. Not everyone can make this work. When it comes to keeping in motion, though, we're only limited by space and our own imaginations.

Use part of your lunch break to take a walk or bike ride. If there are stairs in your office, make a point of using them instead of the elevator; believe it or not, this can actually save you time each day, as the wait for a lift and stopping at each floor before your destination can often negate the speed of an elevator. If you drive to work, park in the farthest space from the office door.

If you must sit, set an alarm to remind yourself to stand up every 30 minutes. There are smartphone apps now that will vibrate to remind you to get up and move if you've been sitting too long. This helps keep your blood flowing and your mind alert.

It can be tough at first to make these kinds of changes. One of the best ways to keep yourself going during this initially challenging period is to purchase a pedometer to wear on your wrist or on your belt to count how many steps you take each day. You can get a simple pedometer for about $10. Many of my patients now wear a Fitbit, Apple Watch, or Nike Fuel Band. While a pedometer now comes with the purchase of a new iPhone, I still use a free iPhone app called Pacer, which ensures I'm always logging my activity.

Just putting on a pedometer can significantly increase the number of steps we take each day. Researchers have found that people average 2,500 extra steps a day when they wear a counter. That's good, because we really need those steps, but it's not enough. I recommend that my patients seek to take at least 10,000 steps each day.

When it comes to this goal, people who live in large cities often have a step up on everyone else. When the activity-tracking company Fitbit looked at data from more than 1 million users, it found New Yorkers were taking an average of 8,807 steps per day. Those were, of course, people who are already consciously monitoring their health, but other studies have consistently shown New Yorkers to be the country's top steppers, regardless of whether they are trying to get fit or not. Not surprisingly, New Yorkers walk to work at a rate that is significantly greater than people from any other part of the country, and that's reflected in their average body weight, too; residents of The Empire State consistently have some of the lowest rates of obesity in the nation.

Even in New York City, though, 10,000 steps can be a rather lofty goal, but it's a target that almost anyone can achieve with the right lifestyle changes. I have patients in their eighties and nineties who regularly log 10,000 steps each day; if they can do it, you can, too.

Many of my patients have found that when they put on a pedometer, many of life's little annoyances became part of a fun game. I've had this experience as well. For example, I no longer find it frustrating to not be able to find a good parking spot; now, I see it as an extra bonus of a few hundred steps. Likewise, as Jane and my kids will attest, I'm really not a fan of housework. After a long day at the hospital, there's almost nothing I like less. But now, at least, I can take some pleasure in the knowledge that when Jane asks me to sweep the kitchen floor, I'll get to add some extra steps to my daily goal. To me, it is one big video game and I am always looking for new ways to get to the "next level."

While I still use my iPhone pedometer every day and enjoy doing so, I don't expect my patients to. Some do, and that's fine, but the real purpose of the device is to simply create some initial awareness. Once you have made the changes necessary to hit at least 10,000 steps, and have made those changes a part of your normal life, then you can drop the pedometer unless you find it helpful to have that constant feedback to keep you moving.

THE HOTEL BUSINESS has been good in Bama County over the past few years. And with so many Chinese coming to Bapan in search of the secrets to health, happiness, and longevity, local innkeeper Xiu Qui has been doing quite well for herself and her family.

So when I first noticed that Xiu Qui was leaving early from the hotel each morning to go work in her family's small farming plot, I was intrigued.

"Have you ever considered hiring someone to work this land?" I asked as she picked up a hoe to start working the soil. "Then you could work more at the hotel."

In reply, she handed me a shovel. "Here," she said. "You are hired."

Xiu Qui and I worked side by side harvesting root vegetables. I stood more than a foot taller than her, probably weighed twice as much as she did, and was at least twenty years younger, but she made the process of turning the dirt look simple, while I labored over every shovel scoop. We then turned to harvesting the green leafy vegetables that were ready to be picked, stooping, bending, grabbing, and pulling, and then doing it all again and again.

As we worked, Xiu Qui told me that there were two reasons why she would never stop working the land.

"There is no promise of what our life will be like in the future," she said. "Once you give up a piece of land, there is no guarantee you can get it back, and having land is at least an assurance that you will have something to eat if times get tough again."

Also, she noted, the visitors to her inn love her freshly picked vegetables.

"And some of them eat a lot more than others," she said, pointing her hoe at me.

When it came time to carry the vegetables back to town, Xiu Qui grabbed an overflowing sack with both hands and, in one swift move, swung it onto her back. A bit of chivalry, and maybe a hint of guilt over the fact that many of those veggies were destined for my plate, took over, and I offered to take the bag from her.

Xiu Qui set down the bag. When I bent to pick it up I immediately regretted my offer to help. It was *really* heavy, and I was embarrassed to find myself groaning as I lifted it and sweating as I carried it on my back to the village.

When we got back, I asked Xiu Qui if she ever got tired of working both on the field *and* in the hotel. She shrugged.

"These are the kinds of questions that the younger people in our village think about," she said. "For me, it has never been a choice."

I thought a lot about those words in the months following that visit. The abundance of choice, in everything from food and cars to clothes and footwear, is one of the great blessings of the modern world, but we're not always *actually* blessed by it. More choices can mean more opportunities to select what is best for us, but it also means having to make more decisions, having to deal with uncertainty when it's not clear if we made the right decisions, and having to live with regret when we know we've made the wrong decision.

And when it comes to exercise, having a choice often means we select *not* to do things that we know are good for us. Taking a job that demands less physical exercise over one that keeps us in motion throughout the day. Taking a car when we really could have walked or biked. Watching television instead of working in the yard. Sitting on the porch when we could have been strolling the neighborhood with a friend.

That's why it is so important to stack the deck in favor of motion. And the best way to do that, by far, is to make motion *fun*.

MARIO HAD ALWAYS assumed his job would keep him healthy. A second grade teacher, he was constantly on his feet, crouching and bending to get near his students, stretching to write on the classroom whiteboard, and climbing to hang art projects on the strings that crisscrossed the room above the children's desks.

"At the end of all of that," he told a group of patients in a healthy lifestyles group I was facilitating, "I'd just go home, plop onto the couch, and watch TV for the rest of the evening. The most exercise I would get was walking to the kitchen to make a microwave dinner."

Over the next hour, our group talked about how to make diet and exercise changes that we stick to. And I used Mario as an example of someone who is already in motion throughout the day, and thus only in need of a little more *strenuous* exercise outside of work.

After the meeting, Mario asked me a question I get from a lot of patients.

"How much should I be running?" he asked.

"I don't know," I said. "How much do you *enjoy* running?"

"I really hate it," he said.

"Then you shouldn't be doing it," I replied.

When people are trying to get into shape, they often default to running. That's both because it's relatively easy to start doing and we've long been told that's what we're *supposed* to do to get into better shape.

But most people don't really like running. Some people assume that they'll enjoy it once they get better at it, but what usually happens is that people who don't enjoy running don't *make it* that far. They quit, even if all they're doing are the short, 10-minute bursts of intensive exercise I prescribe along with all-day, low-intensity motion.

To me, exercise is a lot like food. There's almost nothing in the world I wouldn't try once to see if I like it. If I'm not sure after that first taste, I might try it again. If it's something I've been told often takes a while to develop a palate for, then I might try it a few times. But if I've given it an honest try, and I don't like it, then there's no sense in putting myself through the experience again and again. It's time to move on.

That's why I told Mario what I tell all my patients: "Do the exercises you enjoy doing, because if you don't enjoy it, you won't do it."

Mario told me he'd enjoyed lifting weights when he was a high school football player. "Then try that," I said.

Now, that certainly doesn't mean that all of your efforts to stay in motion and engage in high-intensity exercise have to be things you absolutely *love* doing. I certainly have met people like Makang who relished working in the fields, but most simply did so without much feeling one way or another. The key, really, is that no one from the village had any contempt for fieldwork; it was simply part of their lives.

If this is where you're at with your efforts to maintain motion and get in a bit of high-intensity exercise each day, then you're doing better than a lot of people. Lucky is the person, of course, who *loves* the things they do day in and day out. And I tell my patients that should be their goal. Ultimately, we should all be striving to identify a high-intensity exercise that fills us with joy. After all, once we identify something we really love doing, we're far more likely to continue doing it. And once you do that, the only question is: Why aren't you doing *more* of it?

By the next week, Mario had purchased a secondhand weight bench and some dumbbells from a sports consignment shop near his home. "After I got home each day this week," he told our group, "instead of going right to the couch, I just went down to the basement and lifted as much as I could in 10 minutes."

Remarkably, he said, just that short burst of exercise seemed to stir up his energy, and he didn't want to sit on the couch after that.

I'm pleased to report that, more than a year later, Mario is still teaching, meaning he's staying in motion throughout the day, and that he is still lifting weights intensively after school, meaning he's getting the short bursts of intensive exercise he needs to improve his muscle tone, bone health, and cardiovascular fitness.

That's a tremendous start. Most new exercisers don't make it very long before quitting. You have only to visit a health club on the first weekday morning of January, then return on the last weekday that same month, to know how few people actually maintain exercise-related resolutions.

That's because having success isn't just about the right combination of exercise, and it's not just about finding exercises you enjoy, as Mario did. It's about finding the right balance, and making it part of the rhythm of your life.

And that doesn't just go for exercise. Rhythm is an essential factor in every part of our lives, from birth to death.

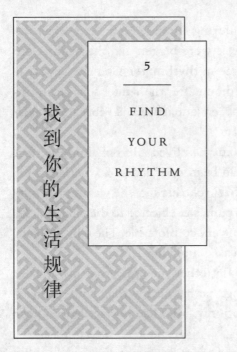

5

FIND

YOUR

RHYTHM

找到你的生活规律

"My life is simple. Because of this, it is easy to know when something is out of balance."

—MAXUE

SHE ROSE EACH DAY WITH THE SUN, AND BY THE TIME IT HAD crested over the lush, green mountains, Maxue had finished her simple breakfast of porridge and vegetables. As the rest of the village began to stir, she would be on her way to the fields. Whenever the rice wines her family vinted were ready, she would carry them to a nearby town to sell.

She ate her midday meal at about the same time each day, and it always consisted of the vegetables, fruit, legumes, or possibly fish that had been gathered and harvested that morning. Every day she returned to work, then returned to sit down for dinner with her family. And every day, as the sun was setting, she went to her loom, where she would weave until the final traces of daylight were gone.

Setting down her crafts, no matter their state at that moment, she would go to her room, lie on a simple wooden mat, close her eyes, and begin to sleep.

The only exception came during Chinese New Year, when the workload was lighter, the meals were bigger, and the nights were longer. Even this, though, came in rhythm; the Chinese New Year comes on the second or third moon after the winter solstice, meaning it is reliably celebrated between January 21 and February 20 each year on the Gregorian calendar.

Age 103 when we met, Maxue had lived a life of almost perfect rhythm. And although she had been confined to a wheelchair for nearly a year following a fall that broke her hip, she was in good spirits. When I asked if she was in pain, she objected to the notion. "Not very much," Maxue said. "I am very comfortable. The only problem is that I cannot move as much as I would like."

When I asked if she had any other medical problems, Maxue laughed. "This is the first thing that has been wrong with me that I can remember," she said. "Until my fall, I had not needed a doctor in my life."

I marveled at this as I checked her pulse. It was strong and steady, and I told her so.

Still, Maxue told me, she sensed she wouldn't be alive much longer. She recalled that her husband's death, in the late 1990s, had come swiftly after years of him being very healthy. And if these were her final days, she said, that was fine. She would rise each day with the sun, as she always had, and make the most of the time she had left. She would greet as many visitors as she could. She would embroider and weave, as she always had. She would live out her time with the same rhythm of life she'd always maintained.

"We are not supposed to be here forever," she said. "It is very good to have a long and healthy life. And when it comes to an end that is good, too. That is part of the rhythm of our lives."

The original bridge into Longevity Village. Until recently, this was the only bridge into the village.

The first thing we saw as we crossed the bridge was a photograph with a biography of each centenarian.

Longevity Soup (center dish), served with a typical meal of fresh vegetables. This soup, composed of hemp seed and pumpkin stems, is eaten regularly and is believed to lengthen one's life.

Centenarian Boxin reaches out to Jane, me, and Elizabeth, inviting us in for a meal.

Jane, Joshua, and Jacob, digging for tubers with our neighbors.

The village's most recent centenarian, Matou, prepares her daily vegetables.

Centenarian Masongmou, recalling how the village kept her alive during the persecution of Mao's Cultural Revolution.

The village coming together to rebuild a neighbor's house.

Jane using her Chinese with the curious elementary school children. The current generation of village children often go home to grandparents and great-grandparents who care for them while the parents work, sometimes in other cities.

The daily commute A village woman makes carrying a fifty-pound basket of vegetables look effortless.

Mrs. Huang, our innkeeper's mother hand farming in Longevity Village. Every vegetable dish she prepared for us wa with vegetables harvested the same day.

Our family shares a walk with centenarian Magan. She can often be seen doing kung fu chops and kicks during these walks.

A village man fishing the Panyang River under the newly constructed bridge. Health seekers come to drink and bathe, believing the river to have healing waters.

Longevity Village's newest residents, escaping the smog and stress-filled cities of China, looking to reclaim their health. Many practice tai chi by the river each morning.

Our family farming with neighboring villagers. With hand farming, these villagers live in rhythm with the natural seasons.

Our family sharing a meal with centenarian Makun. Without any sugar or processed foods available, villagers enjoyed only healthy food choices until recently.

A generation ago this young boy would have had no choice but to live the healthy, traditional lifestyle of the village. Today he faces a changing world around him as automobiles, processed foods, and technology are introduced to the village.

Hundred Demons Cave just outside Longevity Village. During the Chinese revolution, villagers found refuge here from the soldiers who were afraid to enter, believing the cave possessed demons. Now many come from afar, swearing that its geomagnetism and mineral-rich waters heal any ailment.

Longevity Village residents feel their purpose is to nurture and care for family until their passing.

Centenarian Boxin's role, from a young child to his last days, was always to provide for his household. In his early hundreds, when he was no longer able to work in the fields, he gave "longevity blessings" to hundreds of thousands of Chinese health seekers.

Centenarian Magan, working in the kitchen, continuing to make vital contributions to the house and family.

BELOW: Collapse of the original bridge into Longevity Village during our stay in July 2015. The collapse of the bridge represents a turning point for the future of the village.

ABOVE: Da He with his daughter. For centuries, villagers simply lived a healthy life because they knew no other way. With the recent changes in the village, Da He is among the first in his generation to mindfully choose to maintain much of the traditional lifestyle.

Longevity Village at sunrise with the many new tourist hotels and apartments to accommodate the influx of health seekers.

———

AS A CARDIOLOGIST, I have a privileged perspective on the importance of rhythm in our lives. No matter how often I look at a person's heart, be it in surgery or through an echocardiogram, I never cease to be amazed at what this exquisitely designed organ does, day in and day out, from just a few weeks into our fetal gestation until the moment of our death.

To do this so well, and for so long, our hearts must stay in near-perfect harmony with our bodies, speeding up when we need more blood (and the oxygen it carries) and slowing down when that need has run its course. As this happens, our hearts must pump oxygenated blood through the aorta at the same rate at which deoxygenated blood comes in through the superior and inferior vena cava, meaning that the heart's two upper chambers, known as the atria, must beat in near perfect coordination with the two lower chambers, known as the ventricles. This is why, when you listen to your heart through a stethoscope, you hear two beats at a time—*lub dub, lub dub, lub dub.*

Most people's hearts beat about 100,000 times each day. Think of what that means when it comes to reliability. Can you imagine anything that, having been used more than 35 million times in a year, is likely to be just as good at what it does next year as it is right now?

When you think of it that way, it's really quite astounding how *rarely* things go wrong. But sometimes they do. One of the most common problems is when the heart falls out of rhythm and the atria are no longer beating in synchrony with the ventricles. This is called atrial fibrillation, and it represents a significant number of the cases I see in my clinic each day.

About one in four Americans will have an episode of atrial fibrillation before they die. Commonly when this happens, the atria are beating more than 300 times a minute while the ventricles are beating at half that pace or even less. As a result, blood stops moving through our bodies the way it should. Clots can form. That's why, according to research I contributed to in 2011, atrial fibrillation is one of the leading causes of stroke.

Even in lieu of stroke, though, atrial fibrillation can be devastating. It can cause chest pains, shortness of breath, and temporary loss of consciousness. When my colleagues and I looked at the cases of nearly 40,000 patients, we found that atrial fibrillation also dramatically increases the risk of heart failure, premature death, and dementia.

When I started my career, I believed that medications and procedures were the answer to atrial fibrillation. Blood thinners are a common pharmacological weapon in this fight, but come along with no shortage of side effects. For instance, even though we use these medications to prevent strokes, our research shows that they increase the risk of intracranial hemorrhage and possibly even dementia if not used correctly. Another approach to dealing with atrial fibrillation is a procedure known as a catheter ablation, in which radio waves are used to destroy problematic areas of atrial tissue. Even though this is minimally invasive, there are always risks associated with any medical procedure.

Today I approach this challenge in a very different way. Yes, medications and medical procedures can still be necessary, but in my own health journey, and based on what I have seen in China's Longevity Village, I have come to consider these approaches as steps to be taken only when absolutely necessary.

The overwhelming majority of cases, perhaps up to 80 percent, can be prevented, and half of the people with atrial fibrillation can reverse their condition through lifestyle changes aimed at eating better food, maintaining a healthy mind-set, building supportive communities, staying in motion, and learning to connect with their own rhythms.

That might seem like a lot to do all at once, but I like to think of it like riding a bicycle. From moving the handlebars and shifting your body weight, to maintaining momentum and keeping all the moving parts in working order, a lot has to happen to get from here to there on a bike. Yet, once we learn, we don't have to think about it at all. The same is true for the patients who have applied the lessons of Longevity Village to their lives; once they've found their own personal rhythm, they don't have to spend a lot of time thinking about how to keep their atrial fibrillation at bay. They simply have to live their lives.

I SHOULD NOT have been surprised that Maxue's heartbeat was so strong and steady. Having already met and examined several of the other centenarians of Bapan, I'd found them all to be in remarkably good health, especially in terms of their heart health.

Before coming to the village, though, my experience with very aged patients had been different. People who reach their one hundreds usually do so in relatively good health; they pretty much have to be to last so long. But I'm not typically called upon to see healthy people; the folks I see in my practice are always facing heart health issues like atrial fibrillation.

There are about 50,000 centenarians in the United States. Greater than one in four has atrial fibrillation. In Europe, the number is about one in 8. In Bama County, it's one in thirty-four.

Shortly after my first visit to Bapan, I hypothesized that the reason these people's hearts are in rhythm is because their *lives* are in rhythm, and I've yet to find any evidence that contradicts this idea. Just as Maxue's life followed a predictable course from day to day, month to month, and year to year, so too have the lives of all of those in the village who have reached their hundredth year, and many of those from younger generations as well.

That doesn't mean everything was always the same, let alone always good. Far from it. Bapan's villagers have faced many periods of war and political unrest. They have endured periods of dire poverty and experienced, over the past few years, a period of relative fame and economic benefit.

"There is a lot we cannot control," Maxue told me. "All we can do is take care of the simple things in our own lives."

And, as it turns out, that's almost always enough.

Recently, my colleague Jared Bunch and I published a paper about the "tipping point" of atrial fibrillation. In our report, we asked whether it was possible to identify the moment at which changes to the structure of the heart go from reversible to permanent, and while we did identify clear indicators along that path, we also found that it's really tough to answer that question, because there's *so much* going on at once inside our hearts and throughout our bodies.

The same is true for our lives as a whole. Our modern lives are usually a rhythmic mess. We're usually not just out of rhythm in one way, but rather in multiple ways. And while we might be able to withstand a bit of disequilibrium in one part of our life, it's hard to keep our balance when so many parts of our life are so out of sync.

But starting quite simply, we can rebuild this balance piece by piece. And perhaps the best place is where almost all of us start each day, and where we end up each night.

MOST OF US wake up based on when we need to be somewhere, and from day to day that often changes. An early morning meeting can prompt a wake-up that is an hour or two earlier than normal. A late plane flight out of town can mean an extra few hours of slumber in the morning. And even if we keep a typical 9:00-to-5:00 workday, our five-days-on-two-days-off schedules promote sleeping timetables that are anything but routine. For five days in a row we might rise at 6:00 a.m. and go to sleep at 11:00 p.m., then on Friday we might stay up until midnight, 1:00 or 2:00 a.m. and sleep in until 10:00 or 11:00 a.m., only to return to the workweek sleep schedule again on Sunday evening.

To understand how bad this is for us, we have merely to look at the time of the year in which almost everyone is simultaneously thrown off kilter: when most folks across the United States "spring forward" an hour to accommodate for daylight saving time, effectively losing an hour of sleep. On the Monday following the spring daylight saving change, the incidence of heart attacks rises 24 percent, and the impact continues on Tuesday, when rates drop only slightly to 21 percent above the usual rate.

A lack of sleep impacts us right down to the genetic level, affecting the expression of more than 700 genes, which in turn dictate everything from our rates of metabolism, to the way our bodies deal with inflammation, to the antibodies created inside our cells to deal with infections or toxins. Pulling just one late-night work session or just staying up to watch a single TV show leads to the release of some of the same biomarkers that are increased with a concussion.

And we're doing this damage en masse! Up to 70 million of us struggle with sleep according to the federal Centers for Disease Control and Prevention. If you regularly smack the snooze button in the morning, then chances are that you are part of this group. (And any rest you get after hitting the snooze doesn't really even count; the sleep is so fragmented it doesn't do anything for your body.)

Doctors belong to a culture that has long venerated the idea of the sleepless physician selflessly putting patients' needs above his or her own. And sure, the idea of a doctor turning in at 10:00 p.m. each night for a good night's sleep might not be as dramatically alluring as that of a doctor who is staying up late reading through patient records in a heroic quest to save the day. I firmly believe, though, that my commitment to a consistent bedtime when I am home has made me a better doctor for my patients. Indeed, a lack of sleep is the most significant predictor of clinical burnout, a combination of exhaustion and lack of interest in work that looks a lot like depression. You can't help *anyone* when you're burnt out.

Of course, very few of us are in a position to perfectly align our schedules to a sunrise-to-sunset existence. There are, however, things that almost all of us can do to bring a more consistent rhythm to our lives. For starters, we have to address our epidemic cultural habit of defying the sun's power in our lives.

And that's not easy. Many of us have come to identify ourselves as night owls. Indeed, we've convinced ourselves that the hours after the sun goes down are our most productive and our most creative. Oftentimes that is true, although usually not for the reasons we think.

To understand why so many of us feel this way, we have to first remember that all of us exist in a circadian world. Circadian rhythms are biological processes that oscillate every 24 hours, in tempo with a single rotation of our planet. If you've ever marveled at the way a sunflower bows its head at night, then rises to follow the sun across the sky throughout the day, you've witnessed the beauty of this rhythm. The sun is essential, of course, to getting the rhythm going, but once it's started, the plant continues onward even when the sun is hidden behind layers of clouds.

Humans have circadian rhythms, too. But unlike sunflowers, we have the power to tremendously alter the influences that get these rhythms going and keep them in motion inside of us. Even before the sun has gone down, we often have all of the lights in our homes on, and we keep them on late into the evening. Long past the time that our bodies naturally would be shutting down for the night, we're turning on TV shows and working on our computers. According to the National Sleep Foundation, about 90 percent of Americans are exposing themselves to blue light stimuli from computer and TV screens within an hour of trying to go to sleep.

Yet no matter what we do, the biggest circadian stimulus in our lives remains the big yellow ball of hot plasma at the center of our solar system. And this explains why, a million years after we started mucking around with the sun's influence by lighting cave fires, our species remains resolutely diurnal from almost every physiological, evolutionary, and genetic point of view.

As much as we might feel as though we are creatures of the

night, it's simply not true. To prove this, Ken Wright, the director of the Sleep and Chronobiology Laboratory at the University of Colorado, took a group of intrepid volunteers into the Rocky Mountain wilderness with no flashlights or personal electronic devices for a weeklong crash course in circadian rhythm.

Before the trip, the campers, who wore wrist devices to measure the timing and duration of their sleep as well as the intensity of the light to which they were exposed, had all sorts of different sleep schedules. Their levels of the hormone melatonin (which is triggered for release in our bodies as light diminishes naturally throughout the day, thus beginning the gradual process into slumber) were erratic throughout the day and night, particularly among those who were watching TV or working on a computer well past sunset. Within a few days of only being exposed to natural periods of light and darkness, everyone's circadian clocks were ticking together with sleeping and waking periods far more in line with the natural turning of the planet, and hormonal levels far more in sync with one another. Yes, our electronic influencers have a powerful impact on the rhythms of our lives, but our bodies *want* to be in sync with the natural world.

But if we're not supposed to stay up late, then why is it that so many of us feel like night owls? It may be because nothing else is going on; at night, we're left alone with ourselves, with time to think and breathe and get things done without interruption by children, partners, roommates, coworkers, or bosses. Or, we may be mucking up our natural rhythms with unnatural light or putting the wrong foods and drugs into our bodies.

The good news is that we can resolve all of these conflicts. And we can start doing so by simply creating periods during the day that align with the conditions that allow us to actually be at our best when it is best for our well-being. By turning off our phones and shutting down our e-mail. By scheduling time away from the hustle and bustle. By putting ourselves in the mind-set that these minutes or hours are sacredly dedicated to productivity, creativity, or relaxation. It doesn't have to be nighttime for us to make time to be alone with ourselves. It only has to be a priority. And given what it means for our health, happiness, and effectiveness as human beings, it should be.

FOR MOST OF the past century in Bapan, it has been a relatively simple task to be in sync with the sun, because electricity didn't come to the village until just a few decades ago and, even once it did, there weren't many televisions or computers until quite recently. After a long day of soaking up the sun while working in the fields and knowing that there was always another day just like it ahead, villagers had little reason to avoid getting the sleep their bodies need.

The amount of time villagers spend outside is likely a major factor in their longevity. An analysis of studies by researchers at Loyola University has demonstrated a strong link between vitamin D, which we get from the sun, and chronic diseases often associated with aging, including cognitive decline, osteoporosis, cardiovascular disease, diabetes, and cancer.

Not everyone can simply align their lives with the rotation of the globe. Shift workers. Traveling salespeople. Emergency workers. Graveyard convenience store clerks. Hospital staff members. As a cardiologist who is often called upon to care for patients with middle-of-the-night emergencies, this is a fact of life for me, too. There's still plenty that we can do to accommodate for the fact that we live in a 24-7 world, starting with getting as much "sun time" as possible.

Even if it's just for a few minutes, natural light, especially in the morning, is incredibly effective at adjusting our circadian rhythms. That's why I tell my patients to get outside as often as possible. One great way to do this, for many people who work in an office, is to make a "two stop pit stop," a quick stop in a sunny spot any time you leave your workstation to use the restroom. Make sure you're getting sun during other scheduled breaks and mealtimes as well. For people working in large metropolitan areas, this might be the roof of your office building or even that small park a block or two away. Unless there is a cardiac emergency, I make it a matter of habit to step outside after every surgery for a few minutes, giving myself a healthy dose of sunlight and a few moments to decompress before I turn my attention to my next patient.

You should also get as much sun as you can before and after work, and you can make these times doubly beneficial by getting in your daily intensive exercise during these hours as well.

One of my patients, Naomi, is an evening shift manager at a big-box store, where she works each weekday until 1:00 a.m.; it's not the best schedule in the world, but it works for her and her family. When she returns home, Naomi does yoga, takes a shower, says her prayers, and is off to bed by 2:00 a.m. After waking at 7:00 to get the kids off to school, she heads back to bed until 11:00 a.m., at which point she has five good hours of daylight in which to exercise, run errands, pick up the kids from school, and help them with their homework before she needs to be back at work. Although her store is fairly well-lit, it's not the same as being in the sun, so Naomi has made a point of getting an extra dose of light during her break, which she spends sitting by a light box. "It was weird at first," she said. "People were making jokes about sitting under a 'grow light,' but then a bunch of people started asking if they could use it, and now the break room has three of them."

Another patient, Jermaine, worked from 11:00 p.m. to 8:00 a.m. at the airport. Because he and I share the belief that it is best not to get a supersized helping of sunlight before trying to sleep, Jermaine decided to start wearing photochromic lenses that darken as the sun rises in the morning. When his shift ended, he would head home as quickly as possible—no e-mail, no web surfing, no TV—and hit the hay in a basement bedroom with only one small window that was outfitted with "blackout" shades. "It was hard at first," he said, "but

I learned that if I could just keep myself away from any unnecessary light between work and bed, I could be asleep just minutes after I hit the pillow." Seven hours later he would be up and outside, soaking up as much sunlight as he could get and leaving household chores, e-mail, and his favorite TV show for the hours between sunset and the time he had to be at work.

Were Naomi and Jermaine hitting a *perfect* balance of sleep and sun? Absolutely not. Nor are most people in the Western world. But our bodies have a way of thanking us for striving to do what's right.

Even small increases in sleep have been shown to improve our concentration, memory, and mood. In addition to its well-documented impact on age-related diseases, sunlight is associated with stronger immunity, reduced depression, and reduced stress, even in small doses. Fifteen minutes may be all you need to get some vitamin D and keep your circadian rhythm in check. In no way does it take perfection to start reaping these benefits.

DURING HER LAST pacemaker check, one of my patients, Samantha, mentioned that she had been feeling tired all of the time; she was worried it was an indication that her heart wasn't working as it should be. The problem wasn't with her ticker, though, but rather with her coffeemaker.

To counteract the fatigue that comes naturally as a result of living a life out of rhythm, many of my patients turn to caffeine, in one form or another, to give them more energy. While this does provide a short-term energy boost, it often makes the longer-term situation worse. That's because about half of the population has the CYP A12 genetic variation, which makes them slow caffeine metabolizers; any caffeine, even in the morning, can make it difficult for these people to sleep at night. I should know: I am a slow caffeine metabolizer. Even for normal caffeine metabolizers, though, caffeine has a half-life of four to six hours. This means that a good quarter of the caffeine in a lunchtime cup of coffee or soda drink will still be in your system when you naturally should be at least starting to feel as though it's time to hit the hay. And that, in turn, just further throws off our body's circadian rhythm.

It isn't just sleep that is thrown off when slow metabolizers consume too much caffeine. As the caffeine is metabolized, drug levels can build up which may increase the risk of a heart attack by up to 64 percent. The fact that we all metabolize caffeine differently may explain why medical studies seem to "flip-flop" on whether coffee is good for you or not depending on which study you read. Even if you are a slow caffeine metabolizer, studies show that consuming up to 100 mg per day may be safe. To put that in perspective, 100 mg of caffeine is approximately the equivalent of about one cup of coffee, two cups of tea, or four ounces of dark chocolate.

Because high caffeine levels at night make sleep challenging, some of my patients have gotten into the habit of taking a nightcap before bed. But while alcohol can certainly help us fall asleep faster, brainwave pattern studies show that it also heightens both alpha and delta brain activity. Delta activity is connected to restorative sleep, and that's good. Alpha activity, on the other hand, isn't typically present when we sleep; it's much more common when we're awake. The result is a roller-coaster of brain activity that results in restless sleep and leaves us feeling less than our best in the morning.

The people of Longevity Village don't eschew alcohol or caffeine, but their consumption of each is quite moderate. About a third of the centenarians in greater Bama County consumed alcohol during their lives but generally their consumption was limited to small amounts of home-brewed rice wine, similar to what Maxue sold through her family business. No one I spoke to had ever thought to take alcohol as a sleep aid, and drunkenness is all but unheard of. (Part of the reason may be that about 50 percent of Asians have a genetic acetaldehyde dehydrogenase deficiency, which gives them an uncomfortable flushing reaction when they drink alcohol.) When it comes to alcohol and the heart for those who do drink, though, the data are mixed. While it may help to prevent plaque buildup in the arteries of the heart, alcohol is also a known cause of heart failure and arrhythmias.

Rather than coffee, Bapan's villagers may drink tea—not as a wake-up aid, but as a social custom. There is a rich variety of teas in the village, some with caffeine and some without, but in all cases it is consumed in moderation, contemplatively sipped from small cups after mealtimes before the villagers head back into the fields or go out for a post-dinner walk. Although some tea is consumed late in the day, the quantity is very small.

The bottom line for both alcohol and caffeine: If you enjoy them, then only partake in moderation. A single serving of alcohol and one cup of coffee is probably plenty, though. I also advise avoiding any caffeine after noon. According to one study, Americans who maintain a "caffeine curfew" get an average of 42 extra minutes of sleep at night.

NOBODY IN BAPAN had an alarm clock. When we're hitting at least seven hours of sleep on a regular basis, and on a regular schedule, something quite amazing happens: We get the exact amount of sleep our bodies need without having to be jarred awake by a buzzer before we're really ready.

I still *set* my alarm each night. It gives me peace of mind, especially before a morning surgery. But I can't remember the last time I needed it. While my goal is at least seven hours of sleep at night, I don't always make it. My body seems to wake me up naturally when it is time to get up.

That's *real* rest. And the only way we accomplish it is by prioritizing sleep in our lives long enough so that we know what our natural needs are. To accomplish this, I strongly recommend spending at least two weeks shutting down all electronic devices as close to sunset as possible and going to bed at least eight hours before you need to wake up, to give your body plenty of time to adjust to a more natural sleep cycle.

Will you toss and turn the first few nights? Unless you significantly increase your movement on these first few days, I can pretty much guarantee it. In most cases, though, people who complete this exercise for two weeks are able to identify their sleep needs quite accurately, and adjust their bedtimes accordingly. After that, rest simply becomes part of our rhythm, as it should be. And once that happens, we can start to address the other imbalances in our lives.

That's what I told a patient named Jack when I met him shortly after he suffered a heart attack while he was at work.

Jack was forty-three, a bit overweight, had high blood pressure and high cholesterol and was a proud workaholic. The acute myocardial infarction he suffered was not as bad as what I've seen in other patients, but there was damage to his heart tissue that could potentially get a lot worse if not addressed. Complicating matters, his heart attack had caused atrial fibrillation as well. I told Jack it was a warning sign; without significant changes to his lifestyle his heart failure was going to get worse and he would be at risk of another heart attack. If so, it would be necessary for him to get an implantable defibrillator.

"I don't understand," he told me. "I know I don't eat as well as I could, but I ski and I ride mountain bikes a lot."

"How often do you do those things?" I asked.

"I ski once or twice a week during the winter," he said. "And I bike more than that. Probably four or five times a week in the summer."

"Do you go out for scheduled rides?" I asked.

"No," he responded, "just whenever I can get the chance between work."

Jack was a freelance software developer and coder. Sometimes projects would pile up and he'd work for twenty-four hours at a time, sleep for a bit, then get back to work. Other times he'd have several days off to play. When work was crazy, he generally ate at his desk at home, and rarely was it healthy food. When things lightened up, he said, he enjoyed cooking healthy meals for himself and his girlfriend who wasn't around so much when he was busy "on account of the fact that I'm pretty insufferable when I'm working all the time."

"You have no rhythm in your life," I told him. "The way you exercise, the way you work, the way you sleep, and even the way you pursue healthy eating. It's all more sometimes, less other times."

I told Jack what I'd been learning in Longevity Village. We also spoke in depth about medication, procedures, and the possible need for an implantable defibrillator if his heart failure did not resolve. But if we could get his life into rhythm, I told him, his heart condition might resolve itself.

The choice seemed clear. Over the next few months, Jack was a man on a mission. He picked up a stationary bicycle desk, set regular working hours, and carved out time each day to bike ride, ski, or go for a run.

He also set for himself regular mealtimes and started making healthy meals for himself and his girlfriend. That's a more important step than most people think; erratic eating schedules have been shown to result in decreased metabolism, which can lead to long-term weight gain.

It wasn't long, though, before Jack had lapsed into some of his former habits. I suspected I knew why.

"How's your sleep?" I asked him during our next meeting.

"Work keeps piling up," Jack said. "Sometimes I have no choice but to stay up late."

When we discussed this conundrum, I reminded him that it's exceedingly hard to maintain rhythm in one area of our life if we're out of sync in another. I also talked to him about studies linking erratic sleep schedules to heart attacks, brain damage, and dementia.

Jack doubled down. He had to turn down a few projects, but he committed to being done with work each day by 9:00 p.m. and not staying up any later than 11:00 p.m. He also committed to rising with the sun, just like the villagers in Bapan, to get a bright start on everything he needed to do during the day.

Before long, his weight, cholesterol, blood pressure, and resting heart rate had all fallen. His self-reported levels of stress dropped, too. He never needed the implantable defibrillator and his risk of a second heart attack has fallen so far that my recommendation is that he only check in at my clinic once a year. He even reported that his freelance business was going better than ever. He was actually doing *more* while working less; getting more sleep tends to puts us at our best.

The last time I saw Jack, he was like a new person. His heart had completely healed and he no longer suffered from atrial fibrillation. As I entered the examination room I flipped open his file to review his charts.

"Well Doc," he said, "how do I look?"

I looked up at Jack, then back down at the charts.

"You look," I said, "like a Bama County centenarian!"

IF WE'RE ALWAYS starting one thing, then stopping for a moment to do something else, then turning our attention back to the thing we were doing to begin with, it will be more challenging to get our lives into rhythm.

In none of the trips I've taken to Bapan have I ever seen anyone try to multitask at anything. And yet even those who are very old seem to get so much done in a single day. Mawen, for example, would rise each morning to make breakfast, clean up her home, welcome guests, help out with the great-grandchildren, and tend to the livestock. But no matter *what* she was doing, she was only doing one thing.

"That's the best way to get things done," she told me.

One of the great joys of visiting with the people of Longevity Village is the experience of being present with someone as they were

present with me. Even the district mayor, Da Yi, with all of his re-
sponsibilities across several villages, understood that to be most productive he needed to focus on one thing at a time.

"If I am thinking about something else when you are telling me
your problem, then I cannot hope to solve that problem," he explained. "When I focus on one problem at a time, I can solve one
problem at a time. And then the problems don't build up."

Somehow we've come to accept multitasking as an unavoidable
part of our lives in the modern world even though our brains are
designed to only focus on one thing at a time. Multitasking has been
linked to everything from brain shrinkage and lower cognitive func-
tioning to challenges in connecting with others and poor emotional
control. The way I see it, multitasking is simply the art of screwing
up multiple things at once. I'm not alone: There's simply no scientific
basis for the argument that some of us are better when we're doing
more than one thing at a time.

We *are* amazing creatures, capable of performing a wide variety
of skills with expert precision, and yes, even juggling some of those
skills while doing others. Certainly, you can combine a purely physical activity, like walking on a treadmill, and a mental activity, like
answering e-mails, without difficulty. The same could be said of listening to a podcast or a book on your smartphone while working
out at the gym. However, the research is very clear: The more we
combine cognitive tasks, the less adept we are at what's being done.

This might be best exemplified by the words of the fictional physician Charles Emerson Winchester, of the TV series *M*A*S*H*, who
once explained his surgical prowess to his medical comrades like
this: "I do one thing at a time. I do it very well. And then I move on."
Winchester might have been an insufferable snob, but he had the
right approach to getting things done. If we care about doing things
well, we should care enough to do those things one at a time.

Do you think you might be the exception to this rule? Perhaps
because you've got a lot of practice with multitasking? That's what

a lot of people told researchers at Stanford University a few years ago. And when those folks' assumptions about themselves were put to the test, it turned out they were actually *worse* at multitasking than the people who said they didn't have a lot of practice doing more than one thing at a time. They had trouble organizing their thoughts. They got distracted more. They were *much* slower when switching from one required skill to another.

Multitasking doesn't just make us worse at doing things, it might be damaging our brains. People who use more media devices at the same time, like working on a report for work while watching TV, have been found to have less gray matter density in their anterior cingulate cortex, which controls cognition and emotional control.

And since we're talking about living long, healthy, and happy lives, it bears reiterating: Multitasking can be dangerous. For more than a decade, the experimental psychologist David Strayer has been virtually shouting this message from the rooftops at the University of Utah, noting most powerfully that even *talking* on a cell phone can impair drivers as profoundly as drunkenness.

You don't have to be in a car to miss out on what's around you when you are glued to your phone. When the *Wall Street Journal's* Geoffrey Fowler had a friend dress up like Chewbacca from *Star Wars*, then stand waving at pedestrians in San Francisco, those looking down at their phones completely missed the giant Wookiee, even when he was standing right next to them.

Watching a video of this experiment, though, I was struck not by the reaction of those who didn't see the costumed man, but those who did. They smiled. They laughed. They waved. They stopped to take photos. The "smombies" (that's a German-born portmanteau for smartphone-using zombies) missed out on that. And while it might not be a big deal that these people missed seeing a guy in a *Star Wars* costume, it got me to thinking: In a richly beautiful, diverse, and unpredictable city like San Francisco, what else are they missing?

It's not easy to end our addiction to multitasking, but it is possible. Right now, if you're like most people, you have no idea how long it takes you to accomplish even the most routine of your daily tasks, because you're constantly doing more than one thing at a time. It's almost impossible to quit multitasking cold turkey, though, so I recommend identifying a single task that you do every day and committing to do that task without interruption for one week, even if you continue to multitask everything else for the rest of your day.

After a week, you'll have a realistic understanding of how long that task actually takes you to do. You can then schedule time for it accordingly, and move on to another task. This is how we build a better rhythm into our daily lives.

Perhaps the best place to start is e-mail. Almost everyone checks e-mails throughout the day, often alerted to do so by a little "ping" or pop-up window that flashes in the corner of their computer screen. That's a terribly disruptive practice and significantly preventative of establishing any real rhythm throughout the day.

Instead, schedule two e-mail "windows" throughout your day, and make them a bit longer than you think you'll actually need, so that you don't feel stressed by this exercise. Pay close attention to how long it takes you to get through your e-mails during each window and throughout the week. Use some of this time to strategize about how you currently handle e-mail and what you can do to make it more manageable. You can then confidently schedule the number of e-mail windows you need each day. If you overschedule a bit, that's fine; reward yourself with a walk outside or simply a moment to rest and reflect on what you've accomplished and what you're going to do next.

Being present and mindful about whatever it is we are doing, whenever we're doing it, isn't just a good way to be productive. It can also be one of life's most tremendous joys. Instead of enjoying each moment, though, many of us are missing it.

We're losing out on our lives. And that's tragic because, let's face it, even the longest-lived among us isn't going to be here forever.

MAXUE PASSED AWAY in March of 2013 at the age of 104. Like her husband and most other elderly people in this village, her decline had been steep and swift, and her end came peacefully, in no small part because she accepted death as part of the rhythm of life.

Maxue's contentedness with the approaching end of her life is a great example of another aspect of living a balanced existence, one that we often don't like to talk about. Death comes for us all. For some of us it will come sooner. For those of us who choose to embrace a healthier relationship with our lives, it might come quite a bit later. But there is a very big difference between longevity and immortality.

Even though the average American life span has increased by nearly a decade since 1960, our maximum life spans haven't changed much at all. Only one person in history, the remarkable Jeanne Calment of France, is verified to have lived to see her 120th birthday. I'm convinced that some of the centenarians I've come to know in Bama County might live that long, too, but these are one in a billion cases.

But with a healthy lifestyle, it's not at all unreasonable to expect ninety or one hundred exceptionally *healthy* years of life, years in which we will be of sound body, mind, and spirit.

One of the reasons we so often resist living in healthy rhythm with our world is because we don't have a healthy relationship with our own mortality. I'm certainly not suggesting that you should spend long hours dwelling on the impending end of your earthly life, but when we completely *avoid* thinking about death, we are not just avoiding the uncertainty and fear that many of us feel about this part of our lives. We are also not considering the way we wish to approach it.

Perhaps this is why the Bama County tradition of receiving your own coffin at the age of sixty has such profound mental and physical benefits. To the villagers who live with their final resting place in plain sight, day in and day out, for many decades, death is a welcome part of a long and healthy life. Although they have the right to end-of-life care through the hospital in Bama City, very few elders go there when their time comes. Instead they are quite content to die at home, surrounded by the many generations that have followed them. When you are forced to view your own mortality each day it allows you to see the end from the beginning. It brings better clarity. And that certainly has the potential to help put your priorities in order.

I don't think my kids will be buying me a coffin when I turn sixty, but since my first visit to Bapan I've taken to keeping photographs of them when they were very young as the background on my smartphone and computer. Whenever I see these photographs, I'm reminded how quickly time passes. It reminds me to approach life, and death, in the same way.

And for my part, I would very much like to approach it like Maxue. I would like to live a very long, happy, and productive life. I want to be able to ski until my very last day. I want to see my kids grow up and have kids of their own. I want to see my grandchildren have children, too. I don't want to fear growing old—I want to *relish* it.

The real promise of Longevity Village is that if we strike the right balance between our modern lives and some really ancient wisdom, and if we align our personal rhythm to the rhythm of the world, we can better relish the time we have.

We can't do this alone, though. We've already discussed the importance of the people around us. But what I came to learn in Longevity Village is that living a long, happy, and healthy life must be a result of *everything* around us.

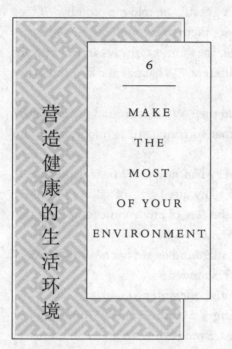

営造健康的生活环境

6

———

MAKE
THE
MOST
OF YOUR
ENVIRONMENT

*"The world is very big; I cannot
change it. But I can make what is
around me better."*

—MAKUN

FOR A LONG TIME, IT WAS A NOVELTY WHEN SOMEONE WOULD
come to Makun's door. Even after her picture was added to the village sign upon her hundredth birthday, back in 2005, unexpected
visitors were quite rare.

Sometime around her 105th birthday, though, things began to
change. More and more people from across China were making pilgrimages to this mystical village, and the days in which Makun was
not entertaining guests became the rarity.

Makun was 108 years old when we first met in 2012 and it immediately struck me that she seemed to be in a perpetual state of
joyfulness. She delighted in telling her story, and the fact that people

170 · The Longevity Plan

sometimes lined up at her door early in the morning, and showed up late into the evening, didn't bother her one bit.

"I have never turned away a guest," she told me proudly. "If they come during a mealtime, I invite them to eat with my family."

No matter when they come, they almost always arrive with the same question, and I was no different. "What has allowed you to live so long?" I asked her.

"It is nothing I did," she told me. "What is around me is healthy, so I am healthy. Everything that surrounds us is important to our health."

At the time that felt like very bad news. While it is possible to emulate in our modern lives many attributes of village life, there is simply no way to re-create the sort of environmental purity that existed in Bapan throughout Makun's life.

"But what if the things that surround us are not healthy?" I asked.

"What sort of things?" she responded.

"Well, for instance, our air and our water," I said.

"But those are just two things," Makun replied. "You cannot be surrounded by only two things. Even if your air and water are very pure, that does not mean you will be healthy. *Everything* around us is important."

When it comes to creating a happier, healthier life, there's almost nothing around us that can't be made better. And since everything around us is important, all we have to do to make the most of our environment is to simply look around. After that, every step we take to de-pollute, de-stress, and de-clutter our lives can be contributive.

First, though, we need to have a better understanding of what's actually putting our immediate and long-term health at greatest risk. And as it turns out, most of us have been getting that all wrong.

FOR ALL THE same reasons the centenarians of Bapan couldn't get sugar, salt, and processed foods for most of their lives, they didn't have access to harsh chemicals, either. As such their ground was not saturated with pesticides, their homes weren't filled with potentially dangerous solvents, and their air wasn't infused with deadly chemicals.

By contrast, most of us in the Western world are *surrounded* by harmful chemicals. In our air. In our food. In the ground. In many of the products we buy and bring into our homes. Unless you grew up long ago in a place like Longevity Village, you were likely raised in a location with some degree of chemical pollution, and you're likely still being impacted by that in some way.

We should all be actively fighting to make the world our children inherit a much better and safer place. But for the purposes of our own health journeys it doesn't do us much good to wring our hands over the fact that we weren't born into a low-chemical community. Rather, it is important to focus on what we *can* do to impact the parts of our environment we actually have some control over.

By far, when it comes to harmful chemicals, the environment we can impact the most is the one in which we live. Yet many of us treat our homes as chemical dumping grounds, usually without even realizing it.

When I asked one of my younger patients what she used to clean her home, for instance, she ticked off a list of all the chemicals and cleaners she kept under her kitchen sink. Then she paused for a moment to think.

"That's just the kitchen," said Tanisha, who is twenty-six and suffers from an inherited rhythm disorder in her heart. "There's also stuff under the bathroom sink. Do you need to know that, too?"

I asked her to continue, and she did.

"OK, so that's the kitchen and the bathroom," she said. "Do you need to know about the stuff in the basement, also?"

The cleaning products she listed were a veritable chemical weapons stockpile. Many of them, while perfectly legal for sale in the United States at this time, have been banned in other nations, emit toxic fumes, and include high concentrations of ingredients that have been linked to asthma, cancer, hormone disruption, and other health problems.

I told Tanisha what I now tell all of my patients: One of the easiest steps we can take to improve our environment is to get rid of a lot of the stuff we've bought over the years on the gravely mistaken notion that it was helping us keep our homes cleaner and healthier.

You can start working toward that goal right now by going to the space under your sink, or wherever else you keep your cleaning products, and tossing out anything that isn't natural. Replace those products right away with lemon juice, white vinegar, rubbing alcohol, baking soda, and borax. Almost everything that can be cleaned with the harsher chemicals we so often use can be brought to a natural state of cleanliness with a combination of these much safer products.

Tanisha was skeptical. "I have two little kids and three dogs," she protested. "You have no idea how dirty my house gets!"

With four kids and a dog of my own, I told Tanisha that I thought I had a pretty good idea. And I acknowledged that, in my experience, these natural products don't disinfect and sanitize as quickly and easily as off-the-shelf products with heavy chemicals.

"But all that dirt," she protested. "All those germs . . ."

". . . are probably pretty good for us," I replied.

I told Tanisha about Makun and the visitors she and the other elders receive each day in their Longevity Village homes. The guests circle around the centenarians, reaching out to touch their hands and faces. They're often hot and sweaty after a long day of travel. And many of those making this pilgrimage aren't in the best of health; for some, of course, that's the very reason for coming to Bapan in the first place.

A few years back, some of Longevity Village's younger residents started to grow concerned about the centenarians' constant exposure to people from all over China. As a doctor, I also had concerns about how these crowds, and even me and my team, might adversely impact the health of the village elders.

When I shared this concern with Makun, however, she waved me off. "I don't worry about getting sick," she said. "I know these people traveled a long way to meet me. If I were to get sick, it certainly wouldn't be intentional."

Whether intentional or not, viruses can be deadly. And they can be particularly deadly for the elderly. As much as I had come to learn that my expectations about health and well-being didn't always apply in this village, I had a hard time shaking off this concern. Any time I saw a large group of Chinese tourists gathered around Makun or another centenarian, I had to control my impulse to run around with a bottle of hand sanitizer.

One morning, a few years after we first met, I learned that Makun wasn't feeling well, and so I headed over to her home to check up on her. When I arrived I was shocked to see her in her usual spot, surrounded by at least ten Chinese tourists, shaking hands, patting heads, posing for photos, and offering words of encouragement. When she saw me standing behind the group she beckoned me over.

"I heard you were not feeling well," I said.

"I'm not," she said. "But these visitors are helping me feel better."

As a smaller group of Chinese tourists arrived, I walked over to the dining area, where Makun's daughter was standing. "If she's already feeling sick her body will not be as capable of warding off other illnesses," I said.

"This is what I have told her," the daughter said. "But she believes the guests bring her strength."

Sure enough, by that afternoon Makun reported she was feeling much better. Her face was flush with color and her eyes were bright. Whenever I'd see her over the next few days, she was energetic and happy.

Makun later told me the worries that visitors might spread illnesses were a concern that younger generations had shared with her a few years back. "Today people seem to think that dirt is bad; it is like they are afraid of it," she said. "But I worked my whole life in the dirt. I would eat wild vegetables and I never cared about the dirt. I have always known that it is good for me."

I THINK IT would be fun to introduce Makun to David Strachan.

Strachan is one of the key minds behind a relatively new line of scientific inquiry often called the hygiene hypothesis. At its most basic, the hypothesis suggests our bodies need some practice battling germs and natural toxins if we're to win the bigger war against allergies and autoimmune diseases. Yet between antibiotics, sanitizers, air conditioners, and simple avoidance of things like dirt, dust, allergens, and pollen, we've created a world in which our bodies simply aren't used to putting up much of a fight at all. So, like a basketball team that doesn't get much practice, we're simply not ready when it comes time for the big game.

Bapan would certainly seem to stand as a point of evidence in favor of this line of thinking.

By the overall sanitation standards of the developing world, Longevity Village is a pretty nice place. The Panyang River carries relatively fresh water right through the heart of the village, and a basic town septic system keeps most human waste away from village activity. People thoroughly cook their food on simple propane ranges. The village's few restaurants appear to be mostly free of vermin and insects.

If you were to invite a health safety inspector from the United States on a tour of the village, though, you can be certain her blood pressure would rise. The river is quite clean by most standards, but there's a 3,000-person town just a few kilometers upstream and even though the Chinese government has been working to prevent pollution into the waters, there are still people who dump their sewage and garbage right into the river. The food here is mostly fresh vegetables, but each day the town's butcher slaughters a pig and parades the carcass around the village on the flatbed trailer of a motor scooter, dicing up smaller chunks on the same cutting board he used the day before with only a cursory washing in between. And when it comes to insects and vermin in restaurants, *mostly free* is great, but that's definitely not the same thing as *free*.

Yet the rate of illness in Bapan is so low that there has never been a permanent doctor in the village. Makun has reminded me each time we have met that she has never needed a doctor. "Not one day in my life," she proudly says.

The hygiene hypothesis is based on a lot of good common sense, and it's backed with a quickly developing body of research. As someone who has seen firsthand what devastation germs and toxins can bring to an otherwise healthy human being, though, I was at first hesitant to agree that we'd gone too far when it comes to cleanliness.

Our twentieth-century obsession with eradicating germs from hospitals, restaurants, schools, and our homes, after all, might be considered one of the greatest achievements of human history. Good hygiene and sanitation have saved, improved, and extended billions of lives over the years. One place we can really see the impact of changes in the way we think about germs is the absolutely precipitous fall in deaths from infectious diseases like pneumonia, influenza, and tuberculosis in the United States in the first eight decades of the last century, owing to better attention to sanitation and germ-killers like penicillin. We haven't yet eradicated these diseases, but we've come remarkably close.

But is it possible many of us have gone too far? That seems likely.

Many modern in-home dishwashers, for instance, are built to reach a National Sanitation Foundation standard that requires a 99.99 percent kill rate for all bacteria. That standard came into question, though, following the highly publicized release of a study showing that children from families who wash dishes by hand are less likely to develop eczema, allergic asthma, and hay fever than kids from families with dishwashers.

Of course that aligns with what we know from Longevity Village. There were no dishwashers there at all until very recently, and even now there are very few. There were also no antibacterial soaps, or really any soap at all for that matter. And to the extent that dishes were washed at all, it was either in the river or in a small bucket of springwater.

Bill Hesselmar, who led the dishwasher study at Sweden's University of Gothenburg, says it's possible that people who don't use dishwashers are socially alike in some *other* way that is preventative of these sorts of sicknesses. Maybe, for instance, they are also more likely to pick up their child's pacifier off the ground, give it a good suck, and pop it back into the little one's waiting mouth. According to an earlier study, also led by Hesselmar, children whose parents do that, rather than having their binkies cleaned with soap and water, were significantly less likely to develop eczema and asthma. Maybe they're also more likely to have a dog at home when their children are infants; that's yet another attribute that studies indicate is preventative of the development of allergies later in life, a finding that struck me as quite interesting as I watched children, some just barely old enough to walk, playing with mangy stray dogs in the streets of Longevity Village.

Clearly, there is a balance to be struck between basic sanitation and complete sterility, and between commonsense cleanliness and over-the-top obsessiveness. As such, these days I advise my patients to strive not for sterility, but rather for a *natural* state of cleanliness. And, if they have to choose, it's probably a good idea to err on the side of nature, in all of its natural dirtiness.

ANYONE WHO HAS ever been to Beijing for more than a day or two understands quite viscerally the immediate damage the Chinese capital's air can wreak on human lungs.

The Chinese government does appear to be waking up to the realization that many of its cities are virtual "airpocalypses." In recent years, hundreds of high-polluting factories have been shuttered or ordered to retrofit their operations to ensure fewer toxic emissions. China still has a very long way to go before its citizens can be assured that they are breathing anything that resembles clean air, though, and in this regard it is like many other countries in the world. According to recent World Health Organization data, one in every eight deaths worldwide is now due to breathing bad air. One study focusing on Chicago, Los Angeles, New York City, St. Paul, and Winston-Salem found that breathing bad air could rob you of up to five years of life.

This is also a tremendous problem where I live. In 2015, the American Lung Association ranked Salt Lake City as the seventh most polluted city in the nation for short-term particle pollution, well ahead of much larger and more industrialized cities like Pittsburgh, Philadelphia, and Indianapolis. In my experience, people are often confused about how this could be until they actually see Salt Lake, which is surrounded on all sides by mountain ranges that top out at more than 10,000 feet. Essentially, we live in a giant bowl, and in the winter months, especially, it generally takes a big storm and a complete flip in pressure systems overhead to flush out the smog that accumulates along the valley floor. It's not unheard of, during these times, for my city to register particle levels higher than those in Beijing.

What can we do about this? What I have learned, and what I tell my patients, is that there are really just three options.

The first option is the least practical for most people: Move.

That's what Qui Yi did. The nurse from Xiangyang, a city of 5.5 million people in China's Hubei Province, had been battling with night restlessness and daytime exhaustion for several years, and was also starting to suffer from irritable bowel syndrome, when she learned that there were very simple homes in Bama County that could be had for less than 20,000 yuan (about $3,200) for a 30-year lease. After retiring from her job and moving to a tiny, dilapidated farm home across the river from Bapan, she resolved to rebuild her life in a place where she could breathe freely.

"In the city the air is always smog," she told me as she worked alongside a neighbor fixing up her old farmhouse with new slat wood walls. "I would never think to open my window for a breeze, so I just suffered in the heat. In the daytime, you could only see ahead of yourself five meters, and the rain would not help at all; the rain was like poison."

In her Bama County home, Qui Yi wasn't just able to open her windows. An entire area of her front wall was open to let the mountain wind breeze through her dwelling.

After just two months in her new home, all of Qui Yi's health conditions were resolved. "Before I could only sleep for two or three hours each night," she said. "Now I sleep for six to eight hours each night, and everything else that was wrong is much better."

Qui Yi's experience makes a lot of scientific sense. By comparing sleep monitoring data from thousands of patients in seven US cities to air pollution information from those same locations, researchers have been able to easily see a strong correlation between bad air and bad sleep. Scientists aren't precisely sure why this is, but we do know that small particles inhaled into our lungs can quickly be passed into our bloodstreams causing inflammation throughout our bodies.

Although it has been more than a decade since the American Heart Association first warned that exposure to air pollution was a significant contributor to cardiovascular disease and death, many of my new patients are still unaware that bad air is one of the major risk factors for heart attacks, arrhythmia, and stroke.

Most people can easily imagine moving from one place to another to pursue a better job or move into a better home. I've found among my patients that it is a lot harder to envision moving for better air. But given the impact bad air can have on our health, and the impact our health has on every single other part of our life, moving to pursue better air shouldn't be dismissed outright, and I ask my patients to at least consider the possibility.

One significant point of resistance, especially since many of my patients have suffered heart attacks in the past, is the idea of being farther away from a hospital. Isn't being closer better? A study conducted in Ohio by sociologists Takashi Yamashita and Suzanne Kunkel seemed to confirm this common belief; it demonstrated a strong association between distance to hospitals and heart disease mortality. When the researchers accounted for socioeconomic and sociodemographic factors, however, the relationship disappeared. It turns out that people who lived closer to hospitals also tend to have higher incomes, which are correlated with better health outcomes.

"If you improve your environment," I often tell my patients, "you're less likely to *need* a hospital."

RETURNING HOME AFTER a long trip to China, a few years back, I looked down from the airplane window into the thick, gray abyss below. It was as if I hadn't left Beijing.

As the plane touched down on a runway so clouded with smog that I couldn't see the terminal until we were right upon it, I looked around at the other passengers. My thoughts turned to my local patients. Tenured teachers. Retirees living on small pensions. Mid-career police officers. These people couldn't just up and move.

Since moving isn't a viable option for many of my patients, I often ask them if they have the capacity to get out of the smog, at least for a while, when it gets bad. For my patients who live in the Salt Lake Valley, where smog levels tend to be worst during the winter months, that can be as easy as taking a hike up any of the nearby canyons or getting to an elevation where the smog isn't as dense. I've found that the psychological impact of this alone can be incredibly restorative, and while the mountain air might be nothing more than a placebo, I really couldn't care less. When something works, it works.

Whether it be skiing, snowboarding, snowshoeing, or some other winter activity, nearly all of my patients who engage in such pursuits have reported positive effects.

One patient, a forty-four-year-old bus driver named Amir, came to me with atherosclerosis, a buildup of plaque on the arterial walls, which can cause heart attacks and strokes, and has been shown to be causally related to air pollution. Among the many interventions I suggested was what I like to call "powder therapy."

"I do love skiing," he told me. "But it's too expensive to go very often."

I jotted down some of the medicines Amir would likely have to take if we couldn't find some more natural interventions, then I asked him to take a look at his insurance coverage and figure out what they'd cost him out of pocket.

A few days later, he gave me a call. "So," he said, "I bought that ski pass."

Skiing wasn't the only intervention Amir needed. We worked together to build a healthier way of eating, exercise, and sleep regimen that would fit his life, in addition to his regular powder therapy sessions. He later told me, though, that getting a doctor's order to hit the slopes "was the best medicine ever!"

Many of my patients, of course, are from other parts of the country. With them, I engage in discussions about options for temporarily escaping the pollution near their homes. Do they have mountains nearby? Forests? An ocean?

About 40 percent of the people in the United States live in a county with a coastal shoreline and, generally speaking, the air near the ocean, which produces about half of the oxygen we breathe, is quite a bit cleaner than it is inland. In many cases, I've learned, a trip to the beach in these counties can be less than an hour away from my patients' homes via mass transit. It can be even faster via personal vehicle, but I encourage everyone, whenever possible, to use the bus or train. If you're trying to alleviate your own exposure to pollution, after all, it doesn't seem right to create more pollution in order to do it.

With more public transit systems offering free wireless Internet, with cheaper and easier access to mobile Internet through smartphones and tablets, and with more people taking advantage of flexible hours and work-from-home options, I have patients on both the West and East Coasts who are engaging in "beach therapy" on a regular basis. One patient who telecommutes in Southern California makes it a point to head to the beach any time there's bad air in her neighborhood; she doesn't dare tell her coworkers that she's often dipping her toes in the surf during conference calls.

Of course, not everyone is in a position to take bad-air-day sojourns to cleaner-air locales. It is for these people, and everyone else for that matter, that a third strategy comes into play.

MOST PEOPLE I know don't have lives or jobs that allow them to leave town whenever the air gets bad. These people are no less deserving of fresh air, of course. For them especially, and really for all of us, it's important to ensure the air in our living and working spaces is as clean as possible.

In fact, this might be the *most* important thing we can do. That's because our indoor air quality is often two to five times worse than whatever we're being exposed to outside. Considering that 90 percent of our time is spent indoors, the quality of our indoor air is of paramount importance.

Firstly, that means getting good air into our homes as often as we can. In most places in the United States the air is pretty clean at least some of the time, if not most of the time, and can be tracked through state and municipal air monitoring agencies. On "green air days" I tell my patients to swing their doors and windows wide open, even if it's cold outside, to take advantage of what nature has to offer.

On bad air days we want to do just the opposite, opening doors and windows as *little* as possible. I also strongly recommend that my patients look into the purchase of high-efficiency particulate arresting air purifiers, often called HEPA filters.

Houseplants can be another great way to keep your indoor air clean and healthy. Back in the late 1980s, around the time scientists were starting to consider getting deeper into space, and maybe even traveling to another planet, NASA produced a report on the ability of some quite common plants to pull chemicals like benzene, trichloroethylene, and formaldehyde from a confined environment, presumably in the hopes of finding ways to keep the air in a long-range spaceship clean. They concluded that very common houseplants like gerbera daisies, peace lilies, dragon tress, and bamboo palms were all quite good at this.

A lot of people think about buying organic when it comes to food, but not when it comes to flooring and furnishings. But these products are also a vital part of the air quality in our homes.

Have you ever been to a carpet store and noticed a pervasive odor? That's the smell of lots and lots of carpets off-gassing, and it's not very good for us. The latex backing and stain-proofing processes used in many modern carpets release chemicals such as 4-phenyl-cyclohexene, toluene, and styrene. These are volatile organic compounds, or VOCs, that can cause headaches, liver damage, kidney problems, and even central nervous system damage. With that in mind, here's a good rule of thumb that really would be common sense if we hadn't lived so long in a world in which we were told that these sorts of chemicals were actually making our lives better: If the carpet you're considering buying has that sickly sour "new" smell, don't buy it.

The same goes for furniture. Many home furnishings have been soaked in flame retardant chemicals. One such retardant, known as Firemaster 550, is used to treat foam, like the cushions in your couch, and has been identified as a common contaminant in household dust, which we're constantly breathing in. Researchers found that rats exposed to Firemaster 550 for just over a week suffered from advanced female puberty, weight gain, and heart enlargement. Most furniture is still chemically treated, but increasingly organic options are available, and they don't *look* organic, they just look like furniture.

For my patients who do have carpets in their homes, I advise being as zealous as possible about keeping them clean. Carpets can capture pollutants from the air and re-release them when walked upon. That's one of many reasons why it's good to follow another principle of many modern Asian homes: When inside, trade your shoes for slippers. A HEPA vacuum cleaner can be a very good investment for homes with carpets, as well.

I also advise my patients to exercise inside on bad air days, doing yoga and aerobics, lifting weights, running on treadmills, or using other home exercise machines. And if you have a choice as to when you're going outside to exercise, the morning hours might be your best bet. In many polluted places, particulate matter tends to spike in the evenings, especially in colder months. Meanwhile, ground-level ozone, which is the product of chemical reactions between nitrogen oxides and volatile organic compounds in the presence of sunlight, tends to spike in the midday through the afternoon.

But while air quality is what so many of us think about when we consider what most pollutes our environment, there is another form of pollution that I've come to believe is just as bad, and which almost no one thinks about.

WE NICKNAMED THE rooster Fred, for his cock-a-doodle-doo sounded more like yabba-dabba-doo, the famous catchphrase of the foolhardy protagonist of *The Flintstones* cartoon series.

Our Fred, one of many roosters in Longevity Village, begins his crows around 4:00 in the morning, which is great if you're the type who likes to get up before the sun. I'm an early riser, but not *that* early, so I just had to get used to him. I wake up to his crow, throw my pillow over my head, and fall back asleep.

Even if Fred wasn't breaking the morning silence, there's not really any such thing as morning silence anyway. When I do wake up in the village, usually around 5:00 a.m., and go for a morning jog, I sometimes stop and just listen to what the world is telling me.

The river is the most constant sound. When it's running high it can be quite loud, so much so that it can be hard to have a conversation if you stand too close to its banks. Even when it's running low, though, it's an ever-present force in the aural environment of the village. Like any place, of course, there also are chirping birds and buzzing insects. And even in these early morning hours, there are always other people out, bundling their farming tools, chopping vegetables for breakfast, or huddled with other early risers to share the village news.

Our natural world is not a noiseless one. Chattering animals, trickling water, and rustling leaves have always combined to create an organic soundtrack for our lives that, scientists say, can average 20 decibels or more, peaking periodically when we engage in conversation throughout the day. Evolutionarily speaking, this is what we're designed to hear and handle, day in and day out, throughout our lives. This is what the village sounds like in the morning.

In most American cities, though, the average background noise borders on levels that can cause permanent hearing damage. Restaurants and bars can be as loud as a gas-powered lawn mower. Retail stores can be as loud as a sustained orchestral crescendo. Even attending a spin class at a gym can expose your ears to 100 decibels and more; that's like standing next to a revving Harley-Davidson.

Almost everyone recognizes the negative impact that air pollution can have on their lives. What many of us have never thought about is the real and pervasive effect of noise pollution.

The simple trappings of modern society, from yard tools and cars to helicopters and passenger jets, have become such a part of many of our lives that we tend to ignore them. Yet we are *hearing* them. Our brains are almost constantly registering recognition of these sounds. In the times in which we appear to be completely incognizant of these noises, it's only because we're already being bombarded with other sensory information, a condition that has been called "inattentional deafness." Whether we recognize it on a constant basis or not, all this noise is taxing our system and adding stress to our lives. That has health consequences.

When we think of the impact of noise on our health, we generally think about hearing damage, and we often associate this with periodic and temporary instances of loud noise exposure, like going to a rock concert. But even noises we generally consider quite common, like busy city traffic, can register at 85 decibels. Long-term exposure to noises at this level can cause permanent damage to our hearing, and hearing loss has been shown to be a significant factor in eventual dementia.

Researchers keep finding connections between noise and other health problems, too. In one study, scientists tracked road noise and hospital admissions over the course of seven years in London, and noticed that when the streets were noisier the hospitals got more crowded, particularly with people suffering from cardiovascular disease and stroke.

The risk seems to spike somewhere around the 50 to 60 decibel level. If that doesn't frighten you, it might be because you don't generally think in terms of decibels. Ideally we wouldn't have to, but our aural environments are anything but ideal.

Determining the decibel level in your home or place of work is quite easy to do; there are a lot of free smartphone apps for this task. A decibel level of 50 is the equivalent of a normal suburb and a level of 60 is the equivalent of a conversation in a restaurant. While occasional exposure to decibel levels of more than 60 likely will not cause us any harm, studies have shown that *consistent* exposure to this level of noise may cause a heart attack or premature death.

The evidence of a connection between noise pollution and ischemic heart disease is concerning enough that it's something I routinely ask my patients about. I even advised one patient, a thirty-four-year-old artist named Amber, to consider moving from her apartment on the corner of two quite busy streets as one of the many lifestyle changes she could make to address a fledgling heart condition.

"It's amazing how much more peaceful my new place is," Amber wrote to me a few months after her move. "I don't think I ever even noticed how much stress the street noise was bringing into my life. I actually thought I'd just blocked it all out."

As is the case for people struggling to deal with air pollution, I recognize that most people aren't in a position to move just because of noise. If you're already considering a move, though, and live in a noisy area, it is definitely something you'll want to add into the decision-making mix. Conversely, if you're in a quieter area and considering a move, you might want to think twice.

Whether or not you can move, the other options I suggested for dealing with air pollution—regularly getting away and making your home a safe haven—apply to noise, too. Since we do spend so much of our time indoors, the decibel levels inside our homes and workplaces are far more important to our short- and long-term health than the noise outside. And this is also an element of our environments that we have a lot of power to control. In one noisy Beijing neighborhood, the addition of sound-insulating windows cut the indoor decibel level from 70 to 35.

Carpets and rugs can also significantly help with indoor noise pollution. Drapes, blinds, and the right appliances and furniture can also lower the decibel level. And, of course, you can *always* turn off that noisy television or radio.

Many of us have gotten very comfortable with wearing hearing protection when attending concerts or sporting events. I strongly advise it in other scenarios, too. I never board a plane without my noise-canceling headphones. Even an hour or two of foam-assisted silence can go a long way toward calming our nerves and healing our bodies of the damages done by so much noise.

As is the case in many other aspects of life, back in Bapan things are changing when it comes to noise pollution. There are still very few vehicles in the village, but there are more of them every time I return. And while the river remains the aural alpha in these parts, there are times in which you cannot hear it over the noise of construction equipment when workers arrive to improve local roads or erect new buildings.

None of this has escaped the attention of Makang. "The only sounds we used to hear when we were working were our tools hitting the ground and our own breath," she told me. "Maybe you might hear someone singing or maybe you might sing. That was it. Today we hear many sounds. That is a sign of progress, but I do miss the peace we had when there was not so much noise."

As though she willed it to be so, it seemed to me that, for just a few moments, every unnatural noise ceased. I listened to the wind. I listened for the birds. And I listened to the river, which was running very high that week. It was quite hot on that afternoon and I longed to go to it.

THERE'S JUST ABOUT nothing better than a tall, cold glass of water. And fortunately, in the developed world, we can have one pretty much whenever we'd like.

Even in Longevity Village clean water was long taken for granted. "I didn't know that there was such a thing as bad water until I was an old man," Boxin once told me. "I thought water was always pure. When I found out otherwise, I felt very lucky."

Because the vast majority of us are also quite lucky in this regard, and since the vast majority of the time we really don't have much cause for concern that our water is acutely dangerous, a lot of us have concluded that we don't have to be concerned at all. Yet many US cities are still reliant on water delivery systems and treatment procedures that have hardly changed since the beginning of the last century. Age-old regulations prioritize taking parasites and bacteria out of our drinking water, but generally fail to address harmful chemicals. And when we take water from our taps and study it, we find that it is sometimes tainted with everything from lead and arsenic to chlorine by-products, antibiotics, antidepressants, and even rocket fuel, which was found as recently as 2003 in the Colorado River, which is a major source of drinking water for Phoenix, Los Angeles, and San Diego, among other cities. More recently, residents of Flint, Michigan, filed a class-action lawsuit against their governor, state, city, and local officials for allegedly ignoring warning signs about dangerously high levels of lead in the city's water system, leading to skin lesions, hair loss, and memory problems, among other medical concerns.

There are many ways in which the water coming from our taps could turn toxic. It could be that the source is already polluted. It could be that treatment facilities are bad. It could be that the pipes, running miles upon miles underground en route to our homes, could be faulty. Any of these situations, and especially a combination of them, could lead us to drink contaminated water.

And it's not just lead we have to worry about. Copper, for instance, has been one of many popular materials for pipes going back to the 1920s. By the 1990s it accounted for more than 80 percent of all water distribution pipes being used in residential construction. When copper breaks down, though, it can cause small amounts of metal to be carried away in the water, right through the tap, into your glass and into your body. A bit of copper isn't bad for most people, but a lot can be problematic, and for people with certain conditions even a small amount can be quite dangerous.

One of my patients, Jim, had initially come to see me about his heart failure and atrial fibrillation. Jim was in his early forties when we first met, and he was determined to change his lifestyle and hopefully reverse his atrial fibrillation without medication, so he joined a support group I was leading at my hospital. One day, after I ticked off a list of potential water contaminants (copper being a common one) and the possible health effects (potentially impairing the systems that clear plaque from building up on our brains) Jim approached me with a concern.

"A lot of my family members have died of Alzheimer's disease," he told me.

"How many is a lot?" I asked.

"Both of my parents," he said.

My colleagues and I have published extensively on the topic of Alzheimer's dementia and cardiovascular disease, and quite a few of the heart patients I see, by virtue of their ages and the connections between these diseases, are suffering from Alzheimer's as well as heart rhythm troubles. I don't have all the answers for them, but I do know that those who feel like they're doing *something* to impact their fate end up beating the odds more than those who are resigned to their roll of the genetic dice.

And there was something that Jim could do, immediately and with very little effort, to take a small bit of control and give him confidence that, little by little with other life changes, he could take even more control over his health destiny. We simply tested his water.

Water testing kits are quite affordable. Jim tested the water in his home with a system purchased for less than $30. That test did show elevated levels of copper, though not quite reaching the 1.3 milligrams per liter that would be considered an "action level" of copper for most people. Nonetheless, given Jim's family history, it seemed silly not to do something, especially since several types of very affordable faucet-attached water filters can almost completely eliminate copper and other heavy metals from tap water.

Is eliminating copper from Jim's tap water going to be enough to completely change his fate? Almost certainly not. He still had underlying heart disease, after all. But that small action is a step. And in my experience, taking lots of small steps toward a health goal is a lot easier and often more effective than taking a big one. Moreover, when big steps *have* to be taken, people who have taken a lot of little ones have more practice. It's like getting a running start.

And fortunately, when it comes to our environment, there are *a lot* of other small steps out there we can take.

BUAN WAS SO excited to renovate his home, trading his traditional stick-and-mud-brick house for the simple cinder block structure that has become increasingly common in rural China over the past decade. When it came time to tear down his old kitchen, though, the Longevity Village farmer simply couldn't. He and his wife loved the old dirt-floored space too much. As a result, their new home is a mix of old and new features, and provides an amazing perspective into the contrasts between modern and traditional life in Bapan.

The old kitchen is dark and cavernous. Light streams in from the cracks in the walls. There are a few pots and pans hanging from the ceiling. There is a brick-lined fire pit and a large wok. On a shelf there is a bag of rice. And that's about it.

Whenever I visit older homes in the village, or those that have mixtures of old and new features, I'm often taken by just how stark and simple everything was in the past, and how little "stuff" the villagers needed to be content.

Our environments don't have to be laden with chemicals to be toxic. When it comes to our health and happiness, *clutter* might be one of the most pernicious things in the world. So many of us, though, don't think twice before welcoming more and more stuff into our homes.

Three-quarters of middle-class residents in Los Angeles can't get their vehicles into their garages because there's too much stuff inside. And the result, according to researchers at USC and UCLA, isn't just a lack of street parking. Cluttered homes are linked to elevated levels of cortisol, a steroid hormone released in our bodies in response to stress. In small doses, cortisol isn't a problem, but when we get too much of it for too long, our bodies can be sent into chaos, breaking down muscle and bone cells, impairing metabolism and brain activity, and weakening our immune systems. That's a *toxic* effect.

The problem isn't just limited to physical stuff. Increasingly, our lives are getting gummed up by digital clutter. There was a time, not so long ago, when it seemed impossible that anyone would ever need a gigabyte of storage space on their home computer. Now many of us carry around smartphones with hundreds of gigs of storage, right in our pockets, and have collections of external hard drives storing terabytes of songs, photographs, and movies. Meanwhile, we're uploading even more stuff on "the cloud," so that we can have access to it any time and anywhere. We can sort through years of e-mails, collaborate with people across the world, and work on projects no matter where we are or what time it is.

Remarkably, we've convinced ourselves that all this digital clutter is helping simplify our lives. In most cases, it's doing nothing of the sort. We're busier than we ever have been. Our to-do lists are longer than they ever have been. Our calendars are as chock-full as they've ever been. This isn't simplicity, and it's not good for us.

Consider the desktop on your computer. Are there files and folders scattered all over the place? What does that do to you when you're trying to find something to drag and drop? Chances are that it's impeding the way your brain works. Neuroscientists at Princeton

University have found that the more visual clutter people are exposed to the less effectively their brains work.

I'm certainly not going to ask you to get rid of all of your stuff. I'm definitely not suggesting that we all take up residence in mud brick homes with dirt floors and nearly empty shelves. As a first and very meaningful step toward detoxifying your environment, though, consider whether every new thing you bring into your home might take the place of one or two things that are already there. Buy a new shirt? Drop a couple old ones off at a local thrift store. Get a new kitchen utensil? Look through the crowded drawer of other utensils and see whether there might be a few that you simply don't ever use. A fascinating meta-example of how this works can be seen in British Columbia, where the provincial government in 2001 declared that for every new regulation added to the books, two old rules would have to be killed; the result was a 40 percent reduction in regulation, even as the province held onto rules that were important to protect health, safety, and the environment.

You can do the same thing with your digital life: Do you really need to hold on to every document you've ever received? Are there songs on your iPhone you haven't listened to in years, or some you don't even like anymore? Kill two for each time you add one, and you'll wind up with only the stuff you really like and really need.

Perhaps the most important place to clear clutter, though, is on our calendars. There may be endless storage on the cloud, but there aren't endless minutes in our days. That's why, with every new opportunity that our family has these days, whether it be a work opportunity or a new sports team for our children, we no longer ask whether we can "squeeze it in." Instead we ask this simple question: What are we willing to take off our schedule to allow room to make this happen? If the answer is "nothing" then we don't do it.

All of this aligns perfectly with ways of the Longevity Village elders. Recall that Masongmou became one of the wealthiest people in the village through hard work and a good dose of business sense.

But being rich never meant having more than one of anything she needed. When she would get a new pot for her kitchen, for example, she would find a poorer villager to give an old one to. That's the village way.

WHAT IS AROUND us is meaningless if we don't consider what is inside of us.

The hygiene hypothesis is a good example of this. It doesn't matter how many germs or allergens are in the environment around us if all that stuff *stays* outside of us. None of that stuff is bad or good for us until it interacts with what is *inside* of us. When you think of things in this way, you quickly can see that it's not the external environment we have to worry about so much as the interaction of it with our internal environment. Indeed, *we* are an environment. And in no place is that clearer than in our gut.

Our gastrointestinal tracts have a role that is at once quite simple and maddeningly complex: transferring food into us, sending its nutrients along on their journeys through the rest of our bodies, and getting rid of the rest.

Vital to this role are the bacteria that live at all points along the way. There are a lot of them, about 100 trillion by some estimates. In fact, there are 10 times more of them in our gut than all of the other cells combined in our bodies. They're also quite diverse. While the human body has about 200 different types of cells, our guts can have as many as 1,000 different strains of bacteria, single-celled organisms which help us eat, digest, absorb nutrients, neutralize toxins, and keep nasty invaders like the "bad bacteria," *clostridium difficile*, from taking over the host organism (that's us!).

Did your parents teach you to be a good host? When it comes to bacteria, we should follow their advice, and do what we would do for any guest. We should make sure they're comfortable. We should feed them healthy food that they enjoy. And, if we're having other guests over at the same time, we should do everything in our power to make sure everyone gets along. When we do these things, our bacteria stay happy. And when our bacteria stay happy, we stay healthy.

Having the right gut environment, also known as gut flora, can help you fight cancer. If we were to sterilize your gut of all bacteria, it would very likely increase the chance that rogue cancer cells could take root and begin spreading, unfettered by cancer-fighting T-cells. That's because when our bodies recognize a problem cell, our bacteria act a bit like a cheerleader for our lymph nodes, encouraging the creation of more T-cells than normal. Without the bacteria, we don't have enough T-cells to fight off the cancer.

There's also compelling evidence that our gut flora can fundamentally change the way our bodies digest what we eat. Certain bacteria can cause us to be better digesters, staying thinner and healthier even if we eat a lot of calories. The fact that Bapan villagers can eat as much as they want without any apparent weight gain or other health consequences is one important indicator of an active gut environment. Indeed, the eldest residents of Bama County have exceptionally diverse gut flora with a healthy population of commensal (generally "good") bacteria, especially escherichia and roseburia. That's in large part because of their high-fiber diet.

What happens in our guts is something of a "food fight" between two popular types of bacteria, bacteroides and firmicutes. Generally speaking, the thin among us tend to have more of bacteroides while overweight folks often have more of the firmicutes. But the composition of our gut flora is incredibly responsive to how we eat and how we live, and just a few meals conducive to the growth of either bacterial type can cause a rapid population explosion of one and decimation of the other.

Animal breeders have known for a very long time that when the creatures they care for are given even small doses of antibiotics, they will gain weight, even if their food intake and exercise patterns remain the same. At least initially, this was taken as a sign that antibiotics made these animals healthier, and it wasn't long before some ranchers realized that investing in drugs could pay dividends when it came time to sell their animals for slaughter.

What we are learning now is that the reason why this happens is because a reduced diversity of commensal gut bacteria, killed off by antibiotics, essentially leaves the tougher-to-kill bad bacteria in charge, causing animals to extract more calories and accumulate more fat from the same diet. We're also learning that antibiotics can have the same effect on us.

As a cardiologist, I've long been concerned by studies showing a significant increase in death from heart problems in patients taking some antibiotics, like azithromycin, due to an interruption in the heart's electrical cycle. Earlier in my career I was convinced that this risk existed just while they were taking the antibiotic. Now, based on more recent research, I worry about the long-term risks of heart disease, obesity, and diabetes among those who have been prescribed antibiotics, especially when they aren't absolutely necessary.

Because the significant lack of heart disease in Bama County has always interested me, antibiotics became a focus of my conversations with Bapan's centenarians a few years ago.

When I asked Makun if she'd ever had an antibiotic in her life she looked back at me with some confusion. "I don't think so," she said. "What is it?"

I got the same answer from many others in Bapan of every age. It was possible, I suppose, that they had been given one and didn't remember, or didn't know what it was, but few of them could recall having been given any sort of medicine.

A small number of villagers did recall having received intravenous treatment in recent years, a favorite remedy of Chinese doctors that is often just a saline solution designed to rehydrate patients who have been sick. Even with all the changes going on in the village today, though, and access to modern doctors in towns not too far from Bapan by bus, I haven't been able to find many younger villagers who have taken antibiotics, either.

I'm certainly not suggesting that antibiotics are bad. We must acknowledge the role these drugs have played in saving innumerable lives since Alexander Fleming's discovery of penicillin in the late 1920s. But it's almost incontestable, at this point, that we have overprescribed penicillin and its myriad progeny pharmaceuticals. We've done this to the point that, when it comes to our long-term health, we're often robbing Peter to pay Paul, sacrificing one part of our overall well-being to impact another in a zero-sum and often negative-sum game. It's also clear that we have ushered in a new era of antibiotic-resistant bacteria. Essentially, in their fight to stay alive against these drugs, the bad bugs have had to muscle up.

That's one of many reasons why I've long been a fan of probiotics, foods like yogurt and many fermented dishes that naturally improve our gut environments by giving a boost to the healthy bacteria struggling to keep the bad guys at bay. Researchers increasingly are showing that probiotics can be a good way to address everything from allergies to high cholesterol to depression.

I was visiting Makun's home recently and discussing fermented foods with a friend when a tour guide from another part of China overheard my Mandarin and sauntered over.

"Nǐmen shuō de shì shénme ne?" he asked. "What are you talking about?"

"I was just telling my friend," I responded, "that in all my time in this village I'd still yet to have tián mǐjiǔ."

"You mean the fermented rice dish?" he said, scrunching up his face. "That's because tián mǐjiǔ isn't really eaten anymore."

Upon hearing this, Makun's daughter began to laugh.

"Don't be foolish!" she told the tourist. "I made some this week!"

The flustered man walked away, shaking his head. A moment later, Makun's daughter brought out a small bowl, opened the cover to reveal what simply looked like a dish of rice, and began spooning out a generous helping to me and my friend. It had a smell that I associated with yeasty bread and which my friend said reminded him of beer. It was a little acidic at first bite and also a bit sweet, with a vinegary aftertaste. I'll admit that it wasn't to my initial liking, but I do think I could develop an appreciation for it, just as I have for the fermented soybeans, known as natto, that I often eat in the morning and which provide a hefty dose of vitamin K2.

Are villagers benefiting from eating food that many Chinese have turned away from? In all the time I have spent in Bapan I have never met a single villager who eats a traditional diet who is suffering from constipation, diarrhea, irritable bowel syndrome, or any other gut malady. That, I've come to believe, is largely a result of the fact that they have always consumed a healthy quantity of fiber-rich foods and probiotics, and have thus far avoided antibiotics. In doing so, they're making sure that their internal environment is every bit as healthy as their external environment.

It's not just antibiotics we need to be concerned about. While it's not popular for doctors to say this, it's time more of us did: We *have* to stop polluting our internal environments with so much medication.

To be clear: If you're on a medication right now, you should not stop taking it without consulting your personal physician. At the same time, if you're not having regular conversations with your doctor about the possibility of coming *off* your prescriptions at some point in the future, then you might be needlessly polluting your body.

I still prescribe medications to patients. Sometimes, after all, medication can be life-saving and life-sustaining. At the same time, I've come to believe that the majority of what we are prescribed should be temporary, and can be if we take the proper steps to address our health problems in other ways.

It's quite sad that so many of us have simply accepted that a daily selection of pills is part of our lives, something we take as a routine part of our day without even giving it much thought. The opposite should be true. Any time we are putting a drug into our body, we should be very aware of what we're doing and why we're doing it, and should be closely monitored by a doctor with whom we are having regular conversations about whether it is still necessary or if there are other steps we can take to alleviate the need for medication. Among my patients who have fully embraced the seven lessons of Longevity Village, the vast majority have been able to get off most or all of their medications.

ONCE, AFTER GIVING a lecture at a Mayo Clinic cardiac conference, a well-respected fellow cardiologist approached me as I was gathering my computer and notes.

"I'm afraid I just don't have the willpower that they have in Longevity Village," he said.

"Willpower?" I asked. "What makes you think they have any more willpower than you? For a very long time, the people in this village didn't need willpower to resist the temptations that destroy our health. They simply lived in an environment that was conducive to health and longevity."

There were no decisions to make, I told him. There was no internal struggle. There were no "shoulds" or "should nots." Their entire environment was conducive to healthy, happy living.

There's a Chinese word that helps describe the way that the people of Longevity Village interact with their environments: *huánjìng*. The first syllable is pronounced a bit like the Spanish name "Juan," but with a rising intonation that the Chinese call *yángpíng*, which often sounds to Western ears like the way our voices rise when posing a declarative question, such as "I am?" The second syllable sounds a little like "Joan," but with an opening sound that is a little softer than an English "J," which Chinese speakers make by placing the tip of their tongue to the back of their bottom teeth, and a falling tone, which the Chinese call *qùshēng,* that sounds a bit like how an English speaker would express dejection.

Of course the point of this book isn't to make you a master of Mandarin. But this word is significant because it defines a concept that has helped me understand why so many people in Bapan have been able to live such long, happy, and healthy lives. In Chinese, *huán* means "a ring" or "a circle." *Jing* means "a border, an area, or a circumstance." *Huánjìng,* then, is everything in someone's circle of experiences.

It's not always easy to build a healthy and happy *huánjìng* in the modern world, but it can be done. Little by little, piece by piece, we can build environments around us that are far better than the ones we find ourselves in today. And once those environments are set, living a Longevity Village lifestyle becomes quite simple.

And the freedom we earn after that, is a big step toward focusing on the final lesson: Living lives full of purpose.

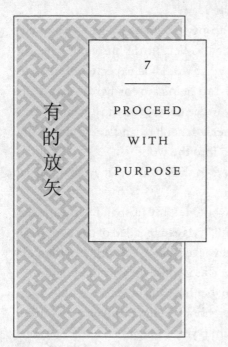

7

PROCEED

WITH

PURPOSE

"My purpose today might be different than yesterday, but I am still important."

—MAGAN

NEARLY EVERY OLDER WOMAN IN LONGEVITY VILLAGE HAS A name that begins with the prefix "Ma," meaning "mother."

While not often practiced today, this was a very old tradition in many parts of China and demonstrates the status and importance of being a mother in Chinese culture. In the communities where this tradition was practiced, a first-time mother would relinquish her own name upon the birth of her child, choosing instead to simply be referred to as that child's mother. This is how the mother of Wen became Mawen. It is how the mother of Kang became Makang. This is how Masongmou, Makun, and the others got their names, too.

But it is not how Magan came to have her name. That is a much longer story, one that helps illustrate the importance of having a purpose in our life, and how that purpose might not be the same today as it is tomorrow. It starts, as many of the village centenarians' stories do, with an arranged marriage.

The girl who would come to be known as Magan was twelve years old when she was matched to her future husband, who was then ten. Even then she knew what would be expected of her once she was wed. She was expected to help care for her husband. She was expected to help his family in the fields. And most of all she was expected to bring blessings to herself, her husband, their family, and the village by bringing children into the world.

They married when they were in their mid-teens. At first, Magan told me, everything seemed fine.

"The people in our village would tell my husband, 'you are very lucky, because she is so beautiful,'" Magan recalled of the early years of her marriage in Bapan. "I never felt like I was prettier than any other woman in the village, but this is what they would tell him."

As the years went by, though, her husband began receiving a very different message from their neighbors. "They would say, 'we are so sorry that you have not had a child yet,'" Magan recalled. "'It must be very sad for you.'"

She was devastated. In China at that time, a woman unable to bear a son was considered unfilial. Under such circumstances, the husband's family had the right to cast her away.

After ten years of marriage, and still with no child, she threw herself before the mercy of her husband and father-in-law. "I should leave so that you can have a family," she told them.

There was no delay in their response. "Why would I ever do that?" her husband said. "I love you as my own child," her father-in-law added. "You are the perfect wife to my son and the perfect daughter to me."

Though heartened by this show of unconditional love, she struggled to accept their decision. Since she was a young girl, she had dreamed of having a family. So, a few weeks later, she came to her husband with another proposal. "Keep me," she said, "but take a second wife so that we can have the family we have always wanted to have."

At those times in China, it was not uncommon for a man to have more than one wife, although the decision to do so was almost always made by the man irrespective of the desires of his spouse. And in Bapan, Magan told me, "there were no families like this because everyone was so poor."

Her husband immediately rejected the proposal, but she continued to work to persuade him. If she could not be a mother, Magan recalled, she would very much like to be a sort of aunt. "I told him, 'I am sure of it. Our family is what matters most. I want to help you find a woman who will bless our family with a child,'" she said.

"Eventually he agreed with me," Magan said. "That is how I got my new sister."

It wasn't long before the second wife had given birth to a baby girl. They named the child Gǎn, meaning "capable."

"My sister said to me, 'you will be her real mother,'" Magan recalled. "And I didn't know how to respond, because I had given up on ever being a mother. But that is how I became Magan. Later my sister had a boy named Xue and that's how she became Maxue."

Throughout her life, Magan would be confronted again and again with situations that would challenge her purpose in life. For a time she served with her husband in Mao Zedong's revolutionary forces. Later, after relinquishing then rediscovering her dream of being a mother, she did what all parents must do, helping her children grow up and watching them leave their home. When this happened, she redirected her life once again as a dedicated farmer and businesswoman selling homemade rice wine. When the Cultural Revolution spread across China, however, her business was shut down and she returned full-time to the fields. When that tumultuous time had passed, she once again returned to making and selling wine. When her husband fell ill in his nineties, she and Maxue dedicated themselves to comforting him in his last days. When he died, she and Maxue lived on together as sisters and business partners, selling colorfully embroidered shoes and woven cloth.

The first year we arrived at the village, we saw the weathered wooden loom which rested in front of their home on the right side of the road, just beyond the Longevity Village entrance. Given its geography within the village, this was the first place many visitors would stop when arriving in Bapan. Maxue and Magan became known as "the centenarian sisters." As the only centenarian who spoke Mandarin, a language she had picked up late in life so that she could do business with out-of-towners and communicate better with her grandchildren, Magan said she cheerfully accepted the role of being an unofficial village ambassador to visitors from across China.

Then, from the time Maxue broke her hip until the day she died in 2013, Magan embraced yet another purpose: Caring for her sister.

"There were many times in my life in which my purpose was challenged or changed," Magan said. "I have always been able to find new purpose, though. I think this is why I have been able to stay alive for so long. We must have a reason to live; if we don't, we won't get anywhere!"

IT HAS BEEN more than sixty-five years since China legally banned the practice of plural marriage. For the most part, arranged marriages are also a thing of the past, and traditional arranged marriages, in which children have no say in who their partner will be, are strictly prohibited. Chinese parents who cannot conceive have more options today, as well, including adoption and what is sometimes called "relinquishment"—the giving of a child to another family so that the child's natural parents don't run afoul of the requirements of China's policies limiting the number of children people may have. Under modern law, Chinese women are equal partners in marriage, and younger generations of Chinese have embraced this equality.

For all of these reasons, Magan's grandchildren and great-grandchildren say they have a hard time understanding the decision she made to encourage her husband to take on another wife. For her part, Magan simply says, "It was a different time."

"Not everyone is going to agree with what you choose to do or what you have done in life," she said. "It's always good to consider the advice of friends and family members, but your life belongs to you."

This conviction gave her strength. At the point I met her, Magan was just about the healthiest and happiest person I could possibly imagine, but I had worried quite a bit about how Magan would fare once Maxue passed away.

The sister centenarians reminded me of some of the older married couples I've come to know through my practice. These patients make their office visits at the same time and often sit in on one another's appointments, sometimes holding hands the entire time. Sometimes, of course, by the time they come to me they've suffered significant cardiovascular damage, and even though it's part of life, it's always a tragedy to see someone's partner pass away. It's not unusual in these circumstances for the surviving partner to die within a matter of months, even if that person's health conditions were being very well managed before their spouse's death.

When we learned that Maxue had passed away upon our next visit to the village we ran to Magan's home. There we found her sitting in her front room surrounded by visitors. When she saw us come in, she stood up and smiled.

Jane went to Magan and they took each other's hands. I ran to their sides to translate.

"I am so sorry about your sister," Jane said, with tears in her eyes.

"Don't cry, don't cry," Magan said. "This is all part of life. I miss her, of course, but there is no need for sadness."

We waited for the crowd to clear and spent the rest of the afternoon talking to Magan, reflecting on her relationship with Maxue and her plans for the future.

"I may be 108 years old," she said, "but I still have many things I'd like to do.

"Now my purpose is to focus on the people who come to visit our village," she said. "They have all come from so far away and I hope that I can make their journey feel worthwhile."

The last time I saw her, Magan was still at it. Almost every day she would greet visitors from across China, sometimes just a few people and other times a whole busload. She believed it was her purpose to inspire every one of her guests to live healthier lives.

"When you eat good food and stay very active," she often told her guests, "you can live to be old and tough like me!"

"How tough is that?" I once asked.

"I could fight off an army," Magan responded. "Come watch me!"

With that she grabbed her old wooden cane from where it was leaning up against the wall and headed for the front stairs of her home. A bit worried by what was unfolding, I glanced back at her grandson to make sure he knew what was happening. The grandson, who was standing in the kitchen area, simply smiled and waved us on.

Magan took the steps quickly but, as she came off the last one, the cane tumbled from her hand and I thought for certain she would topple over. Instinctively, I went to her side, but she righted herself and skipped over to pick up the cane like a kid picking up a rock while playing hopscotch.

"I've got it, I've got it," she said, lifting the stick from the ground. With that she began to march down the street, kicking and chopping her way along the uneven road, sending her jade bracelets spinning around her delicate wrists.

"*Zhè zhēnshi gǔwǔ rénxīn!*" I told her when we finally returned to her home. "That was inspiring!"

"*Hǎo. Wǒ jīntiān de mùdì dádàole,*" she replied. "Good. Today I have done my job."

———

THERE'S LOTS OF evidence that people who feel they have a reason to live are more likely to survive and thrive, even when confronted with tremendous trials and tribulations.

Holocaust survivor Victor Frankl famously observed that it was possible to find meaning in life even in the concentration camps where he was held during World War II. And those who were able to do so, he observed, seemed more likely to survive that horrific experience.

Another story from World War II, less known until it was turned into a bestselling book and later the Hollywood film *"Unbroken,"* is that of Olympic runner and US Army Air Corps bombardier Louis Zamperini, who survived a plane crash, forty-seven days adrift at sea, and two-and-a-half years as a prisoner of war in Japan. Perhaps Zamperini's greatest feat of survival, though, came after he was re-patriated following the war. Haunted by his experiences, he suffered from brutal nightmares and began drinking heavily to cope with his psychological distress. Four years later, at the urging of his wife and with the support of the evangelical preacher Billy Graham, Zamperini was able to find new purpose in his life: forgiveness. In the coming years, he dedicated himself to finding the guards and officials who had been his captors so that he could forgive them, even going so far as to return to Japan, just five years after the war, to embrace some of them. Once he'd found new purpose in life, he told biographers, he was able to stop drinking and his nightmares ceased. He lived to the age of ninety-seven.

The challenges faced by most of us may not ever be turned into a book or a movie, but these battles against enemies like cancer, heart disease, alcoholism, and depression are nonetheless quite formidable, dramatic, and even heroic. Against all of these health conditions and more, purpose is an incredibly powerful equalizer. One study from Rush University Medical Center in Chicago showed that being able to define one's life purpose could help prevent cognitive decline in older age, including Alzheimer's disease, and offers four additional years of life. Studies have found these outcomes exist regardless of age, sex, education, or race. This sort of universal correlation is exceptionally rare; different groups of people generally have differing outcomes in scientific studies. Yet the feeling of purpose seems to be a tonic for everyone quite equally.

The obstacle that many of us face just as equally, though, is adjusting to changes in our lives that challenge our reason for living. That's what Magan dealt with when she couldn't start a family, then again when her children moved away, again when she lost her business, again when her husband died, and once again when Maxue passed away. Time and time again, she had to realign her thinking about her purpose.

That's what my older patients deal with when they lose a longtime partner. Indeed, it's what *everyone* deals with as they go through life in this unpredictable world.

Quite often, the difference between people who survive and those who succumb to their circumstances is that the survivors aren't just able to *find* their purpose once, but also to reassess, redefine, or refocus on their purpose as life goes on.

That was the case for Frank, who came to my clinic after suffering a massive heart attack while at work. Frank had grown up exceptionally poor, put himself through college and law school, and over the past thirty years built up a powerful and successful law firm. In recent years, he'd begun taking complex civil rights cases pro bono, and was preparing one such case for trial when he suffered his heart attack.

Even before we met, he knew I was going to tell him he needed to start taking it easier at work. And he was very resistant to the idea.

"My life *is* my work," he said. "It's what I love and what I'm passionate about. If I am not putting everything I have into it, I don't see that there is a point to living at all."

As is often the case after a myocardial infarction, particularly with patients who will have to make significant life changes to stay alive, Frank seemed depressed and despondent. Depression is three times more common after a heart attack, and it increases the risk of a second heart attack.

If another heart attack happened, I told him, it was likely to kill him, and while there were certainly lifestyle changes that could help reduce the risk of a second incident, we needed to buy him time *immediately*. As he was in imminent risk, I spoke to him about the lifesaving merits of an implantable cardiac defibrillator.

An ICD, as these devices are commonly known, is a device about the size of a table coaster that is implanted under a patient's skin and connected to their heart with thin wires. If the patient's heart stops, the device's battery delivers a powerful electric shock intended to restore the heart to its regular rhythm.

"What's a shock like?" Frank asked.

"Kind of like a grenade going off in your chest," I answered. "But the trade-off for that momentary pain is that you get to live."

Frank seemed unmoved. "I just don't see the point," he said. "If it's my time then it's my time."

When I began my career I was surprised at how common reactions like this are. The feeling of hopelessness that often comes with a heart attack is nothing to take lightly. I also know, though, that patients often come around to the idea of fighting to stay alive.

"Please do me a favor and just think about it," I told him, handing him a pamphlet on what ICD patients can expect. "Give me a call tomorrow and let me know what you decide."

Frank called the next afternoon. "Okay Doc," he said. "Let's do this thing."

When I caught up with Frank recently, I asked him to remember back to the uncertainty he felt on that day and about what changed his mind.

"I went home and I called my son at college," Frank said. "He knew I'd had a heart attack, but I didn't tell him about the ICD. I just talked to him about what was going on in his life. We talked about his classes and the girl he'd been dating since high school. I had been feeling really bad about the idea of slowing down at the firm, but when we were talking something felt like it was changing inside of me. And when he asked me for some advice on whether he should ask his girlfriend to marry him, I realized that I wasn't ready to be done being a father to my boy."

FINDING MEANING IN life makes us more goal-oriented and resilient to the struggles we face. Having a clearly defined purpose can give us willpower we never thought was inside of us. It can bring a level of happiness we have never experienced before. It can relieve the stress that has been crushing us for years. Those are psychological effects that have real physiological impact on the rest of our bodies.

Stress, after all, causes our bodies to release adrenaline, forces our blood pressure up, and prompts our glucose to rise. It impairs sleep and activates inflammation throughout our bodies. A sense of purpose, though, can give us what I like to think of as a 30,000-foot view of our lives. When you see things from that altitude, a lot of your day-to-day struggles stop seeming so important. And when that happens, stress either goes away or is embraced as a means to accomplish a bigger life mission.

Having a strong sense of purpose has also been shown to prevent plaque from building up in our hearts and brains, and keeps blood clots from forming. So far, the best available research suggests this happens because having a clearly defined reason to live stimulates the brain-heart nerve connection, hormones, and immune system, and prevents our platelets from clumping together.

What is it about purpose that is causing all these effects? One big part of it is that it simply makes it easier to make healthy decisions on a day-to-day basis.

Trying to maintain a healthy lifestyle is really difficult, after all, when we're surrounded by people who don't seem to have the same sorts of goals. When well-meaning coworkers bring a box of dough-nuts to work. When it's time to wake up and go for a morning jog. We need a *reason* not to eat that doughnut, not to give in to the temptation to simply sleep through our morning run. These aren't just daily decisions; these are acts of tenacity, strength, and endur-ance.

That's because each decision we make throughout the day saps us of a bit of our strength. A University of Albany psychologist named Mark Muraven demonstrated this when he put a plate of cookies in front of some students and told them, quite simply, that they weren't supposed to eat them. The battle that ensued was epic. When it was all said and done, Muraven tested the students' physical strength and energy and found it had been significantly depleted, by nothing more than having to resist the temptation to eat a cookie.

There was a group, though, that came out of the cookie test much better off than others. Those who had made choices to eat healthy long before they ever even met Muraven didn't struggle at all. There was no temptation, so there was no suffering. Remarkably, they were actually *stronger* and more energized at the end of the test.

Purpose is powerful. That's why, when I took patients through a four-month program intended to help them implement the lesson we discuss in this book into their lives, I made sure to remind them, as often as I could, of their purpose. These patients were all over-weight and suffering from atrial fibrillation. The challenge of over-coming these obstacles is substantial, so every day they would get an e-mail from me asking them to share with me their answer to the question: "Why is your health important to you?" Reminding them of their purpose for getting healthy seems to have been a big help.

Of course, we won't always have someone there to remind us of our purpose, so we need to get in the habit of reminding ourselves.

That's what a patient named Natalie did. Like many survivors of breast cancer who have been through chemotherapy, which can damage heart muscle tissue, Natalie was suffering from cardiomyopathy.

"When I found out I was having heart problems, I felt like screaming," she told me. "It seemed so unfair because I'd already fought so hard to beat the cancer. I didn't feel like I had anything left."

But after coming to accept that she had yet another mountain to climb, the single mother of a twelve-year-old girl refocused on the same thing that had gotten her through cancer and chemo.

"When things would be tough, I had this mantra," Natalie said. "I would just sit and try to get everything else out of my mind and say 'I am Marie's mommy. I am Marie's mommy. I am Marie's mommy.'"

To continue to serve that purpose, both in the immediate and for a long time to come, Natalie knew she needed to overcome this most recent health obstacle. That meant slowing down to prepare and consume healthier foods. It meant finding time each day for a short amount of intensive exercise. It meant adopting an always-in-motion approach to work, which initially left her feeling exhausted at the end of the day. And, perhaps most challengingly, it meant getting seven or eight good hours of sleep a night, meaning she couldn't just set aside things she needed to do after Marie's bedtime.

"Since I'd committed to going to bed not too long after Marie did, I no longer had the time to do things late at night like laundry, making lunches, checking my e-mail, and exercising," Natalie said. "That meant I had to do those things while Marie was awake, and at first it felt like all of our 'her-and-me' time was being stolen away from us."

Marie, however, completely understood.

In a Mother's Day card she made at school, shortly after Natalie began working with me, Marie wrote: "I know sometimes you don't feel like there are enough minutes to spend together, but that's because you are working to make sure we will have many, many, many years together instead."

ON A HOT and humid summer evening, my friend, the local district mayor Da Yi, invited me to take a walk with him to the village where he was born, just a few kilometers up the river from Bapan. "Everyone is getting ready for a wedding," he said. "They will be excited to meet you."

When we arrived, the village was nearly empty. Down by the river, though, there was a tremendous gathering. Men were boiling pig meat in a huge pot over a fire. Women were grinding soybeans into tofu. Children were darting this way and that, singing songs and playing games.

"These are the preparations," he explained. "When someone gets married, everyone works together to get ready."

In speaking to many of the villagers that evening, I heard again and again how much everyone had been looking forward to this event. Not this specific marriage, per se, but to *whatever* the next marriage was. All the villagers knew that, once two people announced their intention to get married, everyone would be invited and everyone would have a role in helping out. Everyone would have a purpose. This was further reflected the next day at the wedding itself, which was a day-long celebration in which, once again, everyone seemed to have a role. There were no professional caterers or decorators; everyone who attended the wedding that day had given of themselves to either prepare for or carry out the festivities.

The way in which everyone pitched in, with no apparent thought as to how their contribution would be rewarded, reminded me of what many of Bapan's centenarians had told me about how homes were built in the past. When someone was in need of a home, people from across the village, and even from across the county, would come with tools, materials, and expertise. There were no thoughts of remuneration.

"Everyone was important," Masongmou once told me. "At any

time you always knew that someone might need your help. That meant at all times you were needed. It is very good to feel needed."

Feeling needed, I have learned, is an essential component to having a life purpose. And that might be one of the key reasons why people who spend a meaningful amount of time volunteering live several years longer than people who don't. If you've ever volunteered for a good cause, you know how fast you become *needed*.

Tragically, a lot of people don't realize they're needed. This is a common feeling for older people, but I've also seen it in my younger patients, especially those who don't have families or aren't close to their families for one reason or another. Many of my patients have told me their jobs make them feel like cogs in a machine. Others have said that while their friends and families would certainly miss them when they are gone, they don't feel like they are actually *needed* by those people.

If you don't feel like there are people in your life, right now, who really depend on you, it's definitely time for a change. Volunteer to coach a youth soccer team. Ask to have a role in your religious services. Offer tutoring in your area of expertise for local high school students. Take photographs for people who cannot afford a wedding photographer. These are the sorts of things we've long been told we should do because they are good for our community. It's important to understand, though, that these things are just as important for ourselves, because they make us feel needed.

Pursuing new career opportunities is another way to feel needed. I recognize that not everyone is in a position to switch jobs but if you feel like a cog right now, you should at the very least be entertaining the idea of finding a new job. Even getting out and exploring the job market a bit can be empowering. After all, when an employer offers you a job, what they are really saying is, "You are wanted and needed!"

ONE OF MY favorite things to do as a doctor is to talk to my patients about their goals. In this way I can be assured that whatever medical advice or treatment I offer them is aligned with their life purpose. Sometimes, I've found, patients are surprised by this. It seems they're simply not used to doctors who have the time to make sure that medical decisions fit into a bigger picture.

When I set about asking my patients for their long-term goals, I found that it varied significantly from person to person, from "seeing my daughter's wedding" and "providing for my family" to "getting my energy back" and "helping more people." When I understand their goals, I find it benefits the quality of the decisions we can make together about their immediate medical care and its impact on their long-term care. This, too, is something I learned in Longevity Village.

"As a farmer, I always knew that what I plant today will not affect me tomorrow or the next day or the day after that," Boxin once told me. "But it will certainly affect me. Eventually it will become very important to my life. So when I would plant, I would always be thinking about the future."

Da He, a villager in his mid-thirties, told me that when he is working in the field he frequently has "conversations" with his future self, a habit he picked up from his parents.

"I imagine that I am working next to myself, and I ask myself questions," he said. "I ask things like, 'What should I be doing on the farm right now to make sure our family is taken care of?' and 'What is the best way to raise the children so that they will grow up to be successful?'"

A focus on the future is an essential part of having purpose. It's also a tremendously important part of our long-term economic well-being, which is something we don't often think about in terms of health, but really should.

One of the reasons why so many social and medical studies are adjusted for socioeconomic factors is that financial well-being is correlated to all sorts of health outcomes. That's logical, since income, especially in the United States, is linked with access to health care, what people eat, social participation, and the opportunity to exercise control over one's own life. Most people understand this to at least some extent, yet they fail to plan for their future financial needs. And that might be a result of the fact that we tend to think of our future selves as strangers.

When people think of themselves in the future, their neural patterns tend to activate in the same way as they do if they were asked to think about someone they do not know. Knowing this, researchers from the Stanford Center on Longevity decided to give some people a chance to interact with their future selves through the use of face-aging software. When they did, they found that people were far more likely to make decisions that provided a long-term benefit over a short-term one.

If you'd like to try this, there are lots of photo-aging apps out there; Merrill Lynch even put one online to encourage people to open retirement savings plans. But you really don't need photo software to get to know your future self in a way that will help you make better decisions today. All you have to do is follow Da He's lead and try to imagine the way your future self would answer questions about your present life. This isn't just a good way to promote better financial choices; it's a way to align all kinds of decisions with a long-term view of our life's purpose.

If you do talk to your future self, consider asking about long-term employment ambitions. That's because the truth is that most of us don't save enough for retirement, so we end up working longer than we planned. That can be problematic, since so many people really don't like their jobs. If we're engaged in work that we find to be meaningful and enjoyable, though, it can actually be a really good thing. All of Longevity Village's centenarians were still putting in a full day of work well into their nineties. Some, like Boxin, worked past their hundredth birthday. I've personally worked side-by-side with people in their 80s and 90s pulling vegetables in Bama County fields.

This absolutely doesn't mean you should keep doing a job you dislike if you don't need to. But if you truly enjoy your work, why quit? Why not just back off a bit to allow for more time for doing other things you love, without eliminating something in your life that keeps you mentally and physically active, helps you feel needed, and contributes to your bank account?

You don't even have to do the same work you once did. One of my friends, Paul, was a college professor for more than forty years before he retired. He now spends a few days a week as a greeter at his local grocery store. "I was tired of teaching," he said, "but I wasn't tired of getting up every day with something to do and people to see."

There's tremendously good evidence that working is good for you. Studies show that people who keep working into their seventies in the United States have better health, better self-sufficiency, and live longer lives. After the age of fifty, the death rate among men who work is half of what it is among men who don't.

There's no real secret here. Work keeps us active. It helps us feel needed. It contributes to our economic well-being. It helps give us purpose. It's what people do in Longevity Village, and it's one of the things I tell my patients they should consider doing, too.

SOMETIMES I WONDER what would have happened if someone had approached me years ago with the advice that I give to my patients today. How would I have responded, for instance, if someone had told me that it might be a good idea to consider working well into the years that many other people retired? Or what if someone had asked me whether my lifestyle was in line with my life's purpose?

At first, I think, I would have indignantly said "Of course it is! I'm trying to create a better life for my wife and children by making sure that our family is completely financially secure, and the more patients I see the more people I can help."

But if asked to further consider how miserable I felt all the time, how quickly my health was failing, how little time I was spending with my family, and how exhausted I often felt, I might have had a different answer.

I was fortunate. At just the right time in my life I was introduced to a place in the world where people understand their purpose and align it to their daily actions. I got to see a place where people's diets, mind-sets, interactions with one another, active lifestyles, and environment exist in perfect harmony with their life purpose. I had mentors, like Magan, to help me see that it's OK for my purpose to change over time and helped me understand how to identify my purpose when it's feeling hard to find. And, of course, I had the love and support of a family who desperately wanted me to feel better about my life, so that I could be an even bigger part of theirs.

In the end, the answers weren't all that complicated. They were, in fact, as simple as the recipe for longevity soup. So sometimes I wonder why it took me so long. Why *didn't* someone come to me all those years ago, to prevent me from so many years of pain?

Maybe it's so I could write this book.

Doing something like this is no small project. So when Jane and I began thinking about whether this was something that *should* be done, I spent a lot of time thinking about whether it was in line with my life purpose.

And it was.

As a husband, it has given me an opportunity to work side-by-side with my wife, as we wrote this book together.

As a father, it has let me put down in writing many of the things I want my children to know about living a great life. I absolutely do plan on living well past my hundredth birthday, of course, but the world is an unpredictable place and none of us, no matter how hard we work to be healthy and long-lived, get the ultimate call on that decision.

As a doctor, I've been given the chance to talk to you, and perhaps to inspire you to live in a way that prevents you from experiencing unnecessary suffering from chronic medical conditions or even premature death. If you eat good food, master your mind-set, build your place in a positive community, stay in motion, find your rhythm, make the most of your environment, and proceed from there with purpose, you're going to have a great life.

My patients have had tremendous success reaching their health and life goals by doing these things. I'm confident you will, too.

But that doesn't mean it's always going to be easy. Making these changes in our modern world can sometimes be a challenge.

And now that's even the case in Longevity Village.

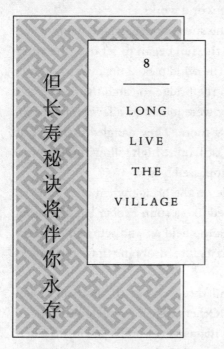

但长寿秘诀将伴你永存

8

———

LONG

LIVE

THE

VILLAGE

"I know my village is special. I want it to always be that way."

—MAGAN

THE FOOTBRIDGE HAD BEEN THERE FOR GENERATIONS UPON generations. Every now and then the wooden slats would rot away, and the villagers would assemble a group of people to replace them. That was all the maintenance that was ever done.

Even after a new concrete bridge was constructed by the Chinese government, a little more than a decade ago, so that cars and trucks could now reach the village, the people of Bapan continued to use the narrow swinging bridge to cross from one side of the Panyang River to the other. Even when it was slick with morning dew. Even during rainstorms. They trusted the bridge to do what it had always done.

I'd never questioned the safety of the bridge, even if I sometimes questioned my own footing upon it. It had always seemed to me that it simply would be there forever.

Of course this is not how the world works.

It was a sticky hot night in the summer of 2015, following one of the hottest days of the year. As the sun began to set over the emerald mountains to the west and the wind picked up, ever so slightly, a few villagers headed down to the bridge to catch the breeze as it cooled over the river. Soon they were joined by a few more people. Then they were joined by a few more. They dangled their feet off the side and watched the water, still quite high following a summer monsoon a few days earlier, as it passed below.

There were dozens of people on the bridge when one of the cables snapped. The bridge twisted like a spun rubber band, flinging people into the waters below. Some held on and scurried to safety. Others dangled for a moment before dropping into the murky brown waters.

The stronger swimmers and those closest to the banks were quickly able to reach the shore. Others were carried downstream.

It took mere seconds for the villagers to assemble, racing to where the river quickens, but shallows, as it forks around an island. Dozens of men and women rushed into the water, splashing and paddling and racing to catch hold of an arm or a leg or a piece of clothing, anything to help those who had been carried away.

Joshua and I were coming back from an evening run when we noticed the commotion. As we approached, I saw Makun's daughter coming up from one of the muddy pathways that lead to the river's edge. She told me what had happened.

"Did everyone get out of the river safely?" I asked.

"I think so," she told me. "But an older man had to be pulled out. He's down there. Would you go see him, Doctor?"

I ran down the road to near Masongmou's home on the riverbank. The man was breathing normally and fully alert. His pulse was fine. He had no broken bones or contusions.

"If you were not all wet," I told the man, "I certainly wouldn't have known that you just went for a swim!"

Twenty-three people went into the river that night. Everyone made it out just fine. Jane had crossed this bridge with Jacob and Elizabeth just moments before it collapsed. We all looked on in awe, grateful that everyone was safe, but lamenting the sad condition in which a historic part of the village now hung.

IT WAS UNCLEAR, as the sun rose the next morning, what would happen to the old bridge, which was still strung across the river on one side but badly twisted and tilted on the other. Village leaders put up a sign, warning people not to cross, but a few men ventured out to examine the damage and a few old farmers, having always taken this route to get to their crops, simply shrugged and went around the barrier, holding on to one of the remaining wires as they stepped, foot before foot, in tightrope style along the way. By that afternoon, a government official had arrived to assess the damage. Everyone agreed that the bridge could be fixed, but the official wondered aloud whether it should. There was, after all, a perfectly good road bridge just upstream. Perhaps the old one should just be torn down.

Later that afternoon, village historian Fu Ji told me that no matter if the bridge was fixed or not it felt as though something had been lost. Almost all of the old homes in the village had been replaced in recent years with simple but more modern cinder block and concrete structures. A village once only accessible by dirt trails was now just a few hundred meters away from a two-lane highway road. On the east side of the village a new "health and wellness center," a large condominium complex being built by a rich developer in an attempt to latch onto the "wellness economy," was in the planning stages. The developer was already taking down payments from people in Beijing, Shanghai, and Hong Kong. Farming plots that had been used for generations were being abandoned as younger villagers did what poor Chinese across the nation have been encouraged by the government to do in recent years, heading into the cities to find work in factories, sweatshops, and on construction crews.

Now the old bridge had broken.

"Everything has changed," Fu Ji said. "And I wonder if this will still be Longevity Village when our children are older."

It had been just over three years since my first eye-opening trip to Longevity Village, but I had seen it and sensed it, too.

Some farmers, who had never known anything but hand tools and animal plows at the time of my first visit, were now hiring contractors with engine-powered equipment to till their plots. Seeds once harvested and stored locally were now being replaced on some farms by giant bags of "super seeds" trucked in from other parts of the country. While pesticides had not yet been introduced to the village the first time I arrived, now and then I would see a local farmer with a big plastic container on her back, walking between rows of vegetables and spraying the crops with chemicals.

There were more local shops, selling trinkets to the tourists, along the village road. There were more motor scooters parked next to old farmhouses. There was more packaged food in the shops. Some parents, when they could afford it, would treat their children to ice cream and soda pop. Even the rice was changing, shiny white increasingly replacing mottled brown.

There were televisions in almost every home, and every year that passed, it seemed, more people were tied to what was on their screens and fewer were concerned with what was happening just outside their front porch.

Health problems hereto unheard of were beginning to creep into the lives of middle-aged villagers. Joint pain and arrhythmic heart conditions. Headaches and digestive troubles. While obesity had never existed in the village before, I was now seeing gradual weight gain among some of our friends in the village.

Yet most people seemed to feel like things were getting *better* in the village, and I was in absolutely no position to argue. These are people, after all, who might have had one or two sets of clothing for most of their lives. These are people who worked arduously long hours, day after day, for decades and decades. And if the footbridge collapse had occurred some years earlier, these are people who would not have been able to get to their crops on the other side of the river, for until just recently that had been the *only* bridge.

It's easy for people who have such privilege to look at places like this and lament what is being lost. The truth is that most people in the village simply don't see it that way right now. In years past, villagers told me, that while they had few material possessions, they did have health and happiness.

THERE ARE A few people in Bapan, and it does seem their numbers might be growing, who have recognized that the very thing that has drawn the attention of people from across China to their village might now be at risk.

Fu Ji and his son are among those who have sounded the alarm. "We don't want to turn back progress," he once told me, "we simply want it to truly be progress."

There are others in Bapan, younger villagers among them, who are recognizing this, too. For now they seem to be a minority, but slowly and thoughtfully they are trying to gather support for changes that would expel some of the worst health intrusions in their village, such as the cigarettes that are now sold in shops throughout the town. By respectfully pointing out that the original plans for the condominium complex exceeded the regional codes for building height, and by encouraging farmers who were going to be displaced to hold out for the best possible price to be bought out of their government land leases, these younger villagers were able to delay the construction of the so-called health and wellness center.

As I watched the very rapid way the village was changing, though, I didn't at first feel confident that those who were trying to save Longevity Village's place as a center of true health and wellness would be successful in doing so. It felt like the village was dying, and like I had arrived just in time for its funeral.

Then I met Da He.

He was wearing his very best shirt, a pink Christian Dior knock-off, along with a pair of khaki shorts and a new pair of shoes he'd purchased for 20 yuan (a little over three US dollars). His three-year-old daughter, Ni La, was tugging at his arm.

"*Zŏu ba!*" she said repeatedly. "Let's go! Let's go!"

Ni La was impatient to see her new baby brother, but the bus would not be coming past the village for a while to come, so Da He was making sure everything in his home was in order for his wife's return from the hospital in Bama City with their infant son.

When I caught him on that morning I knew he wouldn't have a long time to talk, and imagined he might feel rushed, but he pulled two small wooden chairs next to each other and invited me to sit with him.

"*Nǐ hǎo ma?*" I asked. "How are you doing?"

"*Wŏ hěn lèi,*" he replied. "I'm very tired."

He was, though, very excited about bringing his son home. "Now I have both a son and a daughter," he said, "and it feels perfect!"

I asked him if he felt as though his purpose had changed when he became a father. With tears welling in his eyes, he told me it had.

"My purpose is to give my children the kind of life that allows them to be whatever they want to be," he said.

That wasn't a surprising answer, of course. It's the same way parents feel all around the world. But it was refreshing given some of the life choices Da He had made. In recent years, many of the young men and women in Bapan have responded to the pressure villagers from throughout China feel to leave their homes in search of work in the nation's booming cities. Among them were Da He's brother, who had concluded that, in China's fast-evolving economy, money was the best bet for securing a better future for his children.

Da He didn't look down on those choices. Indeed, he said, he often wondered if he shouldn't have followed suit. "But what could be better for my children than their health?" he asked. "What could be better for them than to be able to have a relationship with their father and mother?"

Even among the people in Longevity Village, there are those who say it is no longer possible to live a traditional life. As I have come to know Da He and others like him, however, I have seen that it is indeed still possible to enjoy the health benefits of the old traditions without completely turning one's back on the modern world.

Da He still eats in the way of his ancestors. His diet consists primarily of vegetables, nuts, fruits, grains, and legumes, with regular helpings of sweet potatoes and fish, and an occasional bowl of longevity soup, but no dairy, excessive oils, or refined sugars.

He long ago mastered a positive mind-set. He embraces the simple pleasures and the day-to-day stresses of life. He approaches each day with optimism and playfulness.

Even as many of his friends have moved away for work, he has built a community around himself that supports his life's ambitions. Although he has less money than the people who have gone to the cities, his wife and parents support his decision to stay in Bapan without reservation.

While many people in this region have saved up for motorbikes and motorized farm tools, Da He has continued to walk almost everywhere he goes and farms much as his parents and grandparents did. He's always in motion.

He lives a life of almost perfect rhythm: He wakes when the sun rises, sleeps when it sets, eats at the same time each day and lives each of his days much like the last.

Working the land, he says, reminds him of the power of the environment on his life. This is why he has rejected treating his crops with pesticides or cleaning his home with harsh chemicals.

And he lives his life with purpose, mindful that every new day presents an opportunity to align his actions to his ambitions on behalf of the most important things in his life: his family, his health, and his community.

Da He hasn't rejected all of the trappings of modern life. He lives in a modern Chinese home. He wears modern clothes and shoes. He has a mobile phone and travels when needed, using public transportation. His wife gave birth in a modern Chinese hospital.

"I would not want to live completely in the ways that our ancestors did," he told me. "But I don't think I would want to live completely as many younger people do, either."

His choices seem to be catching on. I have met several others who have gone to the cities only to find that the benefits are not worth the costs, especially when it comes to their health. Some have returned to Bapan. Others have continued to live in the city, but have worked hard to ensure they are eating in the village way, and not picking up habits, like anger and poor sleeping patterns, that seem epidemic outside the village.

It's far too early to know whether villagers like Da He will ultimately prevail by demonstrating that it is possible in Bapan to have the best of both worlds. There's no way to know whether there will still be a reason to call this place Longevity Village a decade or two from now.

But for the time being, at least, the village's reputation seems secure. Despite their obvious overall health, when I first met the seven centenarians of Bapan in 2012, I was quite certain that it wouldn't be long before many of them would be gone. I had simply not wrapped my mind around the idea that people could actually thrive in their hundreds. Five years later, four of our original seven centenarians were still very much alive and well, and new centenarians had been added to their ranks. And those I've met in their seventies, eighties, and nineties, most of whom have continued to live the sort of lifestyle I've described in this book, seem perfectly poised to reach that mark as well.

Will this always be a place where one can come to meet people who have known an entire century of life? I can't say. But I am hopeful, and I believe I have good reason to be.

RECENTLY I WAS walking alone along the Panyang River and stopped to watch a water buffalo crossing against the current. His gray back glistened in the morning sun and he stretched his neck outward, sticking out his thick brown tongue as if pointing it in the direction he wanted to go.

Clouds were wafting between the brilliant green mountains. Across the river a fisherman, wearing a traditional straw *dǒulì* hat, was wading along the bank, pulling a small net behind him. A bit downstream from the fisherman, a mother duck and her ducklings were dancing in and out of a stand of bamboo that was swaying gently in the morning breeze.

"Wow," I said aloud. "I am so lucky to be here."

A wave of gratitude washed over my body and then, quite suddenly, I felt a tinge of guilt, as though I'd been given something I didn't really deserve.

"I wish everyone could have this experience," I said to myself.

It wasn't the water buffalo or the mountains. It wasn't the fisherman or the ducks or the bamboo. It wasn't the sheer majesty and beauty of this place, for there is majesty and beauty in every corner of this Earth.

It was what this place has come to represent for me and so many of my patients. It was that thing I'd set out looking for in what now seems like another lifetime ago.

It was hope.

And everyone *can* have that experience. I have yet to meet anyone who can't improve their health and happiness simply by applying some, if not all, of the practices and principles of the village to their life.

Can we all live to be one hundred or much older than that?

Probably not *all* of us. But I would argue that most people have the potential to reach that or any age in great health. We can get there unfazed by many of the diseases and conditions we've all come to associate with aging.

That's because most cardiovascular disease can be prevented. Many cancers can, too. Most instances of diabetes, dementia, obesity, and arthritis are totally avoidable.

We don't have to slow down. We don't have to give in to the calendar. We don't have to slowly lose our minds or control of our bodies. We can thrive!

And if we can do this at one hundred, then we can sure as heck reach ninety in great health. We can most certainly reach eighty in great health, too. And that means that seventy is going to be just awesome and sixty is going to be simply amazing. And yes, for some of us, one hundred could be just as wonderful as fifty was, or even better.

To have all of those healthy, happy years, the cost isn't that high. It is really just a matter of bringing the seven lessons of Longevity Village into your own life. It's choosing one thing to do to make your life better, succeeding at that one thing, and moving on to another.

And yes, it's totally possible to do this in the Western world.

A long-running study out of Harvard University focused on the lifestyles and health practices of physicians has demonstrated that most people can reach age ninety with great health provided they don't smoke, maintain a healthy weight, exercise regularly, and keep their blood pressure and glucose in the normal range. Great health at ninety really is that simple, and getting those extra years isn't much harder.

As I noted before, greater than nine in ten of the people who have participated in the four-month support groups that test-drove the Longevity Village lifestyle have been able to adhere to their plans and stay on pace to reach their health goals. Most of these people had abused their bodies for years, had decades upon decades of bad health habits, and often had no real support at home. More than a year down the road, most were still living the lifestyle. Most have been able to reverse at least some of their chronic medical conditions, including diabetes, hypertension, obesity, atrial fibrillation, insomnia, fatigue, acid reflux, heart failure, and high cholesterol.

I've done it, too. Despite all the bad genes. Despite my addiction to sugary treats and high-fat processed and prepared foods. Despite all those years of denial in which I told myself I was healthy, even as the evidence mounted that I was falling apart. Despite the damage I'd already done, I was able to change my life.

And you can, too. Whatever experiences you may have had in the past that have led you to believe that living in harmony with the way we were designed to live is not possible, trust that this is actually the simpler way. And over time, it just gets easier. It is indeed the best of both worlds. It's living with all of the benefits of the modern world while still maintaining a connection with ancient wisdom to guide you toward a happier and healthier life.

The people of Bapan have given us an amazing path to follow. Despite unfavorable genetic profiles that should put them at greater risk for a myriad of cardiovascular conditions, their lifestyles have destroyed any notions of genetic destiny. Despite lives that included some immense hardships, they've remained resolutely happy.

And in recent years they have begun sharing their stories far and wide.

"Look around," Magan recently told me. "All of the people who come here are learning something about how to make their lives better. Then they are taking that with them where they go. No, Longevity Village is not dying. Our village is growing bigger every day."

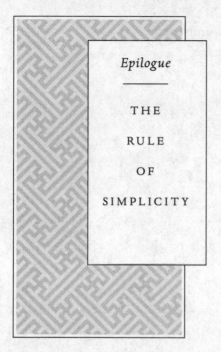

Epilogue

THE

RULE

OF

SIMPLICITY

ON THE EVENING THAT FENG CHUN TAUGHT ME HOW TO MAKE longevity soup, I returned to my room feeling a bit disappointed. I'd always told myself that I didn't believe in magic or health miracles, but I suppose I'd been looking for a secret, and in my mind that secret had to be complicated.

It took a while—a bit longer than it should have, perhaps—before I realized that the secret had been there all along. Again and again I'd seen it.

It was in the soup, yes, but it was in everything else, too.

The elders of China's Longevity Village didn't need studies, exercise regimens, dieticians, or doctors to tell them how to live. They just lived. They didn't even really think about how they were living because, for the most part, they didn't have a choice. They ate longevity soup because it was what was available and they liked it. That's all and nothing more.

The secret, it turned out, was simplicity.

Of course it was. What did I expect?

ACKNOWLEDGMENTS

THIS BOOK WOULD NEVER HAVE COME TO BE WITHOUT THE KEY contributions of many people. It was only after the urging of my wife, Jane, that we made the first of many research journeys to Longevity Village as a family. She has led by example, striving to live and teach the principles of Longevity Village in our own home even before our journeys there began.

I wish to thank the physicians, nurses, and other staff members at my hospital who encouraged me to write this book after hearing me present our findings at our hospital's grand rounds. Never in my wildest dreams would I ever have imagined writing a book.

It was Drs. Charles Mallender and Dan Nadeau who pointed me in the right direction after my colleagues convinced me to write this book. Special thanks to Lone Jensen who took our experiences in Longevity Village and translated them into a book proposal. Thanks to Trena Keating for believing in this book, guiding us along the way, as well as her many hours spent revising early versions.

I wish to thank Gail Winston and her team at HarperCollins for also believing in the message of this book and providing us a way to share our message of hope. Her vision and editorial insights helped us create a much better book than we could have done on our own.

Thanks to Jennifer Nelson and Alecia Anderson who organized the patient Longevity Plan Study and helped so many of our patients adopt the health lessons of Longevity Village to turn their lives around. Thanks to Darryll Gilliland for making possible all of the community Longevity Plan seminars.

Special thanks to Matt Walker and Brennan Snow, from the Tiger in a Jar video team, who documented our second research trip to Longevity Village on film. Thanks to Lv Zeng, my Chinese language coach, who helped me maintain my Chinese language skills to do this research, made arrangements for our research team, helped with interviews, and organized the analysis of the genetic samples.

Thanks to Drs. Yunlong Xia, Yan He, and Leon Ma in China for helping me with the research behind this book.

I am especially indebted to my collaborating writer, Matthew LaPlante, for getting the thoughts, ideas, and experiences out of my head and down on paper. He is an insightful thinker, an incredibly talented writer, and a good friend. Thanks as well to Matthew's wife and daughter, Heidi and Mia, for their support of this project and journeying with us to Longevity Village.

I could not have done this without the lessons learned from my patients. It was our joint struggle with medical challenges that led me on this journey in the first place. Your love, support, and eagerness to learn about this village motivated me through the difficult periods.

Lastly, thanks to our children and my parents. It was special to share many research trips to Longevity Village and the transformational journey with you. Thank you Dad for believing in me and the message of this book.

INTRODUCTION

8 **That's what happened in 2012** Translating the findings of the Bama Centenarian Study into English ignited my passion and desire to understand and experience this longevity belt of China. This 1996 study by Zhenyu Xiao and colleagues was published in the *Chinese Journal of Population Science*.

10 **My first trip** Our team grew in future years to include my language coach, Zheng Lv, who serves as a fixer for our expeditions; Dr. Leon Ma, a cardiologist from Dalian; Dr. Yan He, a professor of cardiology at the Guangxi University Medical School; videographers Matt Walker and Brennan Snow of Tiger in a Jar Productions; and my other two children Jacob and Elizabeth, who along with Joshua and Jane were always willing to jump in to lend a hand to make sure that the medical, scientific, and information-gathering parts of our trips were successful. My co-writer, Matthew LaPlante, along with his wife Heidi and their daughter Mia, visited the village with us in the summer of 2015.

12 **The man, who introduced himself** We had no shortage of people who were willing to help translate between Mandarin and the Bapan village dialect, but recognized by our second visit to the village that it was much easier to have a single translator working to help us communicate with the elders who could not speak Mandarin.

13 **At a reported age** For all of the reasons I've described, there is simply

no way to be 100 percent certain of everyone's exact ages. I have taken the centenarians, their families, the Chinese government's census, and national ID cards, and the many Chinese researchers at their word for the purpose of reporting ages in this book, but I'm far less concerned with calendar years than how healthy these individuals and their communities are.

16 **the average ratio** This is according to the 2010 US Census. For Okinawa, Japan, the centenarian ratio was reported by Craig Wilcox and colleagues from the Okinawa Centenarian Study.

16 **First, there is an extensive body of medical literature** The researchers in the Bama Centenarian Study used additional methods for verifying ages beyond that which was reported in China's census or the national identification cards. Their age verification methods were described in the previously referred to 1996 study by Zhenyu Xiao and colleagues which was published in the *Chinese Journal of Population Science*.

19 **the rate of heart disease in the United States** These data are from the 1998 publication in the *American Journal of Cardiology* by T. Colin Campbell and colleagues in the Cornell China Study.

19 **The rate of breast cancer** In a fascinating study by Xufeng Fei and colleagues as published in *PLOS ONE* in 2015, Chinese researchers showed that the rate of breast cancer was much lower in rural areas of China. These breast cancer statistics stand in stark contrast to the very high levels of breast cancer as reported by the US Centers for Disease Control, which can be reviewed at www.cdc.gov/cancer /breast/statistics/state.htm

19 **The rate of dementia** In a study of 267 Bama County residents over the age of eighty, Chinese researchers could only find one case of dementia. This Chinese language study was published by Wu Yeguang and colleagues in the June 2005 issue of the *Chinese Journal of Behavioral Medicine*.

22 **Our findings are not** The genes of people who reside in Bama County should theoretically have a "modest unfavorable impact," yet they don't, according to a study published by Ning-Yuan Chen and colleagues in the journal *Lipids in Health and Disease* in 2014. In that study, researchers were unable to explain how a so-called "bad gene" might not have the effect that would normally be expected,

and theorized there might be "other protective genotypes" among the aged test subjects. Another answer, of course, is that the expression of this gene was limited by the sorts of lifestyles detailed by Zhenyu Xiao and colleagues who, writing for the *Chinese Journal of Population Science* in 1996, interviewed and examined 69 people over the age of 100 from this region. All of these findings correlate to my experiences in Bapan, with the possible exception of the findings on dementia. I've seen some signs of dementia among a small number of centenarians, but it's by far the exception. Most of the older residents of Longevity Village are fully cognizant, conversant, and even quick-witted.

22 **In one study** This 2014 publication of 516 people from Bama, China, by Ning-Yuan Chen and colleagues in the journal *Lipids in Health and Disease* showed that, based on genetics alone, the people of Bama should have much higher rates of heart disease.

22 **As it turns out** Inheritance, authored by geneticist Sharon Moalem (and co-written by my co-author Matthew D. LaPlante) is a great primer on genetic expression and what it means for our lives. I recommend it to many of my patients who are seeking to understand how genes change our lives and our lives can change our genes.

22 **We've all known** Twin studies are exceptionally helpful for understanding of genes, since monozygotic twins (those we often call "identical") start with the same genetic material. The study I reference here, published by Anne Maria Herskind and colleagues in *Human Genetics* in March 1996, comprised a tremendous number of non-emigrant like-sex twin pairs born in Denmark during the period 1870–1900.

CHAPTER 1: EAT GOOD FOOD

27 **"Secret? There is no** Few Chinese believe in a monotheistic god, and Boxin is no exception. Most, however, believe in an afterlife and worship their ancestors. When Chinese people talk about "God" it is often as part of an expression of the inability to explain why something happened.

31 **Yet many people** If you had to classify the Bama County diet, you would find it very similar to the Okinawa diet or the Mediterranean

diet in that fruits, vegetables, nuts, seeds, and fish make up the majority of the food consumed. Certainly, the medical data supporting these eating habits are impressive in promoting a long and healthy life free from heart disease, cancer, and dementia. The health benefits of the Mediterranean diet on cardiovascular disease, for instance, were explained by Ramón Estruch and colleagues in the April 4, 2013, issue of the *New England Journal of Medicine*.

35 **While there has been** One study of whole grains, tremendous in size and length, followed more than 350,000 participants from 1995 until 2009. It found that people who eat whole grains had reduced risks of cancer, cardiovascular disease, diabetes, respiratory disease, infections, and death from other causes. The study's results were published by Tao Huang and colleagues in BioMed Central Medicine in March 2015.

36 **For those seeking** The food pyramid, developed by the USDA Center for Nutrition Policy and Promotion, shows a range of servings for each of the five major food groups. For grains, the pyramid suggests 6 to 11 servings each day—even the low end of that suggestion is far higher than what people eat in Longevity Village, and this is one of the key ways I disagree with the pyramid's suggestions.

37 **The Longevity Village diet** A study published by Hung N. Luu and colleagues in the *JAMA Internal Medicine* medical journal in May 2015 concluded that peanuts, given their general affordability, may be considered a cost-effective "nut" for the purpose of improving cardiovascular health. That effect was observed in a large study of Seventh-day Adventists in California, published by Joan Sabaté in the *American Journal of Clinical Nutrition* in September 1999, in which the frequency of nut consumption was connected to a substantial reduction in myocardial infarction and death from ischemic heart disease. One reason peanuts and pumpkin seeds are so good at preventing heart disease may be their high levels of magnesium; according to a study published by Stephanie E. Chiuve and colleagues in *The American Journal of Clinical Nutrition*, higher plasma concentrations of dietary magnesium were associated with a 41 percent lower risk of sudden cardiac death.

39 **And then there are the peppers** You don't need to go overboard on the chilis. Scientists who studied the effects of capsaicin, led by

Mary-Jon Ludy for a report in the journal *Chemical Senses* published in February 2012 recommend a moderate level of pepper consumption.

41 **There's certainly no lack** The journalist Brittany Shoot made a compelling case for water buffalo dairy in a March 24, 2014, article for *Modern Farmer* magazine. I'm inclined to believe that if the people in Longevity Village were not lactose intolerant, they would have incorporated a moderate amount of buffalo milk into their diets.

42 **What about calcium?** The highest hip fracture rates are in North America and Europe, where dairy reigns supreme, according to a report by Dinesh K. Dhanwal and colleagues in the January–March 2011 edition of the *Indian Journal of Orthopaedics*.

43 **As in the case** The evidence of a long-suspected connection between dietary saturated fats and coronary artery disease appears to be weak. More recent research indicates dairy products and coconut oil can have a beneficial impact on health, according to a research review published by Glen D. Lawrence in *Advanced Nutrition* in May 2013.

44 **The prevalence of these fish** I haven't always liked fish. After reading the article in the April 2, 2013, issue of the *Archives of Internal Medicine* by my former classmate, Dariush Mozaffarian, though, as well as other studies showing the health benefits of fish, I forced myself to start eating it. Slowly, over time, I began to enjoy the taste. Now I love it.

44 **There are plenty of other** The Monterey Bay Aquarium's Super Green list is intended to help consumers make good decisions that positively impact both human and ocean health. Contributing to the list are scientists from the Harvard School of Public Health and Environmental Defense Fund. The list can be found at www .seafoodwatch.org. One thing to remember when considering the fish on this list is that while many people avoid fish over concerns about mercury, selenium binds mercury and helps to get it out of our body. Ocean fish are highest in selenium. A November 2010 article by Nicholas V. Ralston and colleagues in the journal *Toxicology* details selenium's protective effects against methylmercury toxicity.

45 **"Oh no," Boxin told me** In his book, *The China Study*, T. Colin Campbell very strongly suggests that everyone should eat a vegan diet. Because of the nature of his book, which I highly recommended, some readers might have concluded that rural Chinese are all vegans. Nothing could be further from the truth. As noted, they'll pretty much eat anything that's edible. Whether or not they should eat that way—or that you and I should eat that way—is a very different question.

45 **Before you pass off** These and other recipes for offal (which rather unfortunately is pronounced as a close homophone of the word "awful,") are available on Chris Cosentino's website, www.offal-good.com.

46 **When confronted about** How does one determine that protein was a major driver of human evolution? A research team led by Manuel Domínguez-Rodrigo, an archaeologist at Complutense University in Madrid, came to this conclusion based on a fascinating study of a 1.5 million-year-old child's skull. The study was published in *PLOS ONE* in October 2012.

49 **While 3,400 mg of sodium** Too much dietary salt has long been considered problematic, but recent research suggests too little is a problem, too, as detailed in a report published by the Institute of Medicine of the National Academies in May 2013 and Kirsten Bibbins-Domingo in the January 2014 issue of *JAMA Internal Medicine*.

50 **Fruit juices are** Not surprisingly, those who drink juices that have added sugars are even more likely to gain weight, according to a study published by Dariush Mozaffarian and colleagues in the June 29, 2011, issue of the *New England Journal of Medicine*.

51 **Of course, all of this** The use of saccharin and other artificial sweeteners can change how we metabolize glucose, leaving our bodies less efficient at doing so. This study was published by Jotham Suez and colleagues in the October 9, 2014, issue of *Nature*. Speaking to the journal *Nature* of the results of a study that had one such finding, New York University microbiologist Martin Blaser said the results were "counter-intuitive—no one expected it because it never occurred to them to look."

52 **Water is something** The United Nations has prepared a very good

primer on water insecurity—quick and worthwhile reading for those of us interested in making sure we are taking heed of our blessings and privileges. It can be found at www.unwater.org.

52 **But if we simply** This study, published by Michael Boschmann and colleagues in the *Journal of Clinical Endocrinology and Metabolism* in December 2003, looked at the effect of drinking water on adipose tissue metabolism.

53 **There are no rules** Like everything in life, you can always have too much of a good thing. Hydration is one thing. Drowning our bodies in water is another. As long as we're not gulping down water on a regular basis, though, we can responsibly drink without risking dilutional hyponatremia, also known as water intoxication.

53 **The second guideline** Richard D. Mattes, who published a paper on the difficulties of measuring and predicting human thirst and hunger in the journal *Physiology and Behavior* in April 2010, notes "the health consequences of drinking in moderate excess of need are minimal." To me that means that if you think you want something to eat, particularly during non-meal times, try a glass of water first. There's no reason not to.

54 **Hacking the deep-brain impulse** The effect seems to work best on older patients and those who are obese, according to a study published by Emily L. Van Walleghen and colleagues in the journal *Obesity* in 2007.

59 **When you eat** One of the most frequently cited studies on this effect was published by Pedro Rada and colleagues in the journal *Neuroscience* in 2005.

61 **How big of a fight** These findings are interesting, but the risks can easily be overstated. Also, the research design and reporting process needed much improvement. *Washington Post* reporter Stephanie Pappas did a good job of separating the actual scientific takeaways from the headlines in a piece she published on October 21, 2013. The bottom line: The study doesn't prove Oreos are as addictive as cocaine, but it does raise interesting further research possibilities.

63 **Fresh vegetables and fruits** Diane M. Barrett, the director of the Center for Excellence in Fruit and Vegetable Quality at the University of California, Davis, produced a great literature review on this subject in *Food Technology* magazine in April 2007.

63 **Even more troubling** "The dominant effort," explained the study's lead author, Donald Davis, "is for higher yields. Emerging evidence suggests that when you select for yield, crops grow bigger and faster, but they don't necessarily have the ability to make or uptake nutrients at the same, faster rate."

65 **Regardless of where** One particularly telling study by Dariush Mozaffarian and colleagues showing the relationship between vegetable and fruit intake and weight loss was published in the *New England Journal of Medicine* in the June 23, 2011, issue.

68 **Is there a time** We also found, however, that excess levels of vitamin D could be connected to a substantially increased risk of atrial fibrillation. The lead author on our study was Megan B. Smith and the abstract was published in the November 22, 2011, issue of *Circulation*.

69 **I came to learn** Ben Horne's work includes a paper, published by the *American Journal of Cardiology* on June 1, 2012, demonstrating that routine periodic fasting was associated with a lower prevalence of diabetes. His findings in that paper built upon animal studies, like the one published in April 2004 in the Proceedings of the National Academies of Science of the United States of America showing that caloric restriction—taking in fewer calories while avoiding malnutrition—slows the development of age-related diseases.

70 **It doesn't stop** If you're interested in diving deeper into the subject of intermittent fasting, a research review by Giusi Taormina and Mario Mirisola published in *Biomolecular Concepts* in April 2015 is a very good place to start. It's important to note, as medical journalist Roger Collier did in the June 2013 edition of the *Canadian Medical Association Journal* that intermittent fasting as a diet strategy might be "moving into fad territory." That can cause two very different problems. First, Collier wrote, some doctors "are prone to dismissing fads out of hand," thus preventing their patients from legitimate benefits. Second, many promoters will seek to cash in by offering simple and ultimately flawed advice that results in extreme behaviors, such as fasting for two days each week and binge eating during the rest.

74 **Oftentimes, when researchers** A study published in the *Asia Pacific Journal of Clinical Nutrition* in 2004 by Irene Darmadi-Blackberry and

colleagues didn't bother with any sort of nuance when it came to the connection between beans and health. It was called "Legumes: the most important dietary predictor of survival in older people of different ethnicities."

74 **What we tend** Cheese lovers went crazy over a study, published by Hong Zheng and colleagues in the *Journal of Agricultural and Food Chemistry* in March 2015, that purported to show that cheese was a key to the so-called "French paradox," which was an observation that French people suffer from relatively low levels of coronary heart disease even though they have a diet that often includes lots of saturated fats. One important caveat: the study included just 15 young, healthy male participants. The fromage jury is still out.

CHAPTER 2: MASTER YOUR MIND-SET

77 **The duo had been singing** Arranged marriages have been banned in China since 1950, but the laws governing marriage in China have become more exacting over time.

78 **But when Mao** From 1966 to 1976, Mao orchestrated one of the most disastrous political and social disruptions in modern history. Enforced by the Red Guard—which could be every bit as intimidating and terrorizing as its name intimates—the revolution included attacks on artists and intellectuals, political purges, and economic disaster.

79 **No one lives to be so old** In one study of centenarians that is worth some introspective contemplation for us all, Peter Martin and coworkers from the Georgia Centenarian Study shared the five personality traits often seen in centenarians. In particular, people who live over the age of one hundred exhibit low levels of neuroticism and high levels of competence. Their findings were published in the December 28, 2006, issue of *Age*.

79 **I've become convinced** Matthew J. Hertenstein and colleagues, in the June 2009 issue of the journal *Motivation and Emotion*, theorized that people who appear to be more happy in childhood photos might gravitate toward others with similarly sunny dispositions.

81 **There is an old parable** One of the earliest, if not the earliest, English versions of this story can be found in *Zen Action: Zen Person*

by Thomas Kasulis of the Department of Comparative Studies and Department of East Asian Languages and Literatures at the Ohio State University.

82 **Buddhism teaches** This is one of the Four Noble Truths, as interpreted by the fourteenth Dalai Lama. As is the case with any ancient religious teachings, there are lots of different interpretations of the Dhammacakkappavattana Sutta, which is a record of the first teaching given by the Buddha, but most seem to agree that craving for things, both that we have and do not have, is the origin of suffering.

89 **There are tremendous** In a study of about 1,500 Dutch workers published in *Applied Research in Quality of Life* in March 5, 2010, Jeroen Nawijn and coworkers demonstrated that those who took vacations reported being quite happy at work in the period leading up to a vacation. After a vacation, their level of happiness at work was about the same as people who took no vacation at all. Similar impacts were observed in a study, published in the *Journal of Applied Physiology* in October of 2007, that showed changes in cardiovascular function when people were anticipating sleep.

89 **In a study** Becca R. Levy and colleagues found that people with more positive perceptions of aging, measured up to 23 years earlier, lived 7.5 years longer than those with more negative perceptions, according to an article published in the *Journal of Personality and Social Psychology* in July 2002. Scientists still aren't sure why negativity has such an impact on our genes, but they do know there's a causal relationship.

89 **Telomeres are** Change in telomere length over time has turned out to be a good measure of biological age. Elissa S. Epel and coworkers writing for the *Proceedings of the National Academies of Sciences in the United States of America* in December 2004, for instance, demonstrated that people who face severe emotional stress, in particular, are most likely to have more rapid deterioration in their telomeres.

90 **With all that in mind** One of the best studies demonstrating the relationship between stress and bionic age was published by Janice K. Kiecolt-Glasera and colleagues in the journal *Brain, Behavior, and Immunity* in May 2010.

90 **Studies show that** A University of Wisconsin study of 28,753 people showed that whether or not stress was healthy really came down to our perception of that stress. Indeed, those who embraced stress lived 17 percent longer according to lead author Abiola Keller in a September 2012 article in *Health Psychology*. https://www.ncbi.nlm.nih.gov/pmc/articles/PMC3374921/

90 **Only about 13 percent** These figures are from a 2014 Gallup poll. While the 46.7-hour average workweek didn't represent a tremendous jump over years past, it was still higher than at any time since 2002, according to Jena McGregor of the *Washington Post*.

91 **Back in 1988** While it's hard to make an apples-to-apples comparison between smoking and unsupportive work environments, I don't make this claim lightly. The study, called "Work-Based Predictors of Mortality: A 20-Year Follow-Up of Healthy Employees" by Arie Shirom and coworkers was published in the May 2011 issue of the journal *Health Psychology*.

93 **My friend** Rachel Lampert and colleagues, writing for the *Journal of the American College of Cardiology* in October 2014, explained a study involving nearly 100 patients who kept a journal of their daily emotions for a year. Their study demonstrated that negative emotions such as anger, anxiety, and sadness can put people at greater risk of atrial fibrillation, thus adding AF to a list of conditions (including ventricular arrhythmias and myocardial infarction) that can be negatively impacted by these sorts of emotions.

95 **Li Yu probably** The researchers, Norbert Schwarz and Spike W. S. Lee, (the latter is the social psychologist from Hong Kong, not the film director from New York) reviewed their work and the work of others in a paper published in the October 2011 edition of *Current Directions in Psychological Science*. "Cleaning one's hands removes more than physical contaminants," they wrote, "it also removes residues of the past."

97 **There is a lot of natural evidence that play** This is a fascinating study by Robert and Johanna Fagen showing that those bears who played the most were the most likely to thrive and survive. These findings were reported in *Evolutionary Ecology Research*, 2009.

99 **A lot of the things** The study from which this data came was a secondary data analysis of a 933-person cohort of novice runners

from Denmark. It was published by Rasmus Oestergaard Nielsen and team in *PLOS ONE* in June of 2014.

99 **A lot of medications** This study, published by Takehiro Sugiyama and team in the *Journal of the American Medical Association Internal Medicine* in July 2014, isn't the only place where prescription drugs are thought to be prompting poorer choices on the part of patients. I have seen the same thing with diabetes and high blood pressure medications as well. Too often, once people start on these medications they start making poor food choices again. They feel a false sense of security on the medications.

102 **Some of the most** If you overlap two maps of the United States— one of heart disease rates and the other of angry tweets—the correlation is absolutely stunning. The research was published by Johannes C. Eichstaedt and colleagues in the February 26, 2015, edition of *Psychological Science*.

104 **Most people** The article "Respiratory Feedback in the Generation of Emotion" was published by Pierre Philippot and coworkers in the journal *Cognition & Emotion* in 2002.

105 **We can use** An interesting report on the evolution of "anger face" was published by Aaron Sell and team in the journal *Evolution and Human Behavior* in September 2014. It argues that the muscle movements that constitute the human facial expression of anger are inherited—they exist even in blind children who have never seen an angry expression—and evolved because they increased others' assessment of an angry person's strength.

106 **An important thing** The findings of a meta-analysis published by Marcus Mund and Kristin Mitte in the September 2013 edition of *Health Psychology* have been really overstated in the popular media—particularly the connections drawn to expressing anger and longevity, which need to be further explored, but the analysis is very compelling when it comes to demonstrating differences in levels of hypertension between those who express emotion and those called "repressive copers."

CHAPTER 3: BUILD YOUR PLACE IN A POSITIVE COMMUNITY

111 **There's a powerful** The law, called "The law of protection of rights and interests of the aged" was passed in 2012 by the Chinese legislature. It does not specify the penalties nor define the legislature's definition of the word "frequently."

112 **It would be very hard** The research by Patricia A. Boyle and colleagues included 823 older individuals who were free of dementia when the study began. During follow-up, 76 of the test subjects had developed Alzheimer's disease. Those who initially described themselves as lonely were compared to those who initially described themselves as not lonely. The study, "Respiratory Feedback in the Generation of Emotion," was published by Pierre Philippot and coworkers in the journal *Cognition & Emotion* in 2002.

112 **Yet even as** These millions and billions are a bargain, really, considering the costs of caring for smoking-related illnesses and excessive drinking. One study published by Xin Xu and coworkers in the *American Journal of Preventative Medicine* in March 2015 pegged the cost of getting someone to stop smoking at just $480 per quitter.

112 **Perhaps it's no wonder** I'm not generally inclined to look at our social history with rose-colored glasses, but at least in this respect, the evidence is damningly sad. The shift seen by Miller McPherson and team was most pronounced when it came to a loss of neighbors as close confidants, according to a study published in the June 2006 edition of *American Sociological Review*.

113 **As it turns out** Julianne Holt-Lunstad and colleagues describe the very real mortality risks from loneliness and social isolation in a meta-analysis study published in *Perspectives on Psychological Science* in the March 2015 issue.

116 **First, they asked** This second study by Julianne Holt-Lunstad and coworkers required volunteers to keep a journal about their interactions with people they had contact with. The researchers later compared the diary entries to the blood pressure measurements taken before, during, and after each contact. The results were published in the July 2003 edition of *Health Psychology*.

117 **Why not just sprint** A study published by Rena Wing and Robert

Jeffery in the February 1999 edition of the *Journal of Consulting and Clinical Psychology* showed that people who went through a weight-loss program alone only maintained their weight loss about a quarter of the time, while those who went through the same program with three friends and family members maintained their full weight loss two-thirds of the time.

118 **When a social psychologist** We may remember negative messages, but the messages that inspire us to change our behaviors are positive ones, according to Tracy Epton, Paschal Sheeran, and their colleagues' analysis.

119 **According to the USDA's** The NPD Group, a leading global information company, found that consumption behaviors in the United States have become less household-oriented and more individualized than previous generations, owing in no small part to a rise in the number of households that consist of just one person.

119 **These aren't positive trends** The results of the study, published by Annalijn I. Conklin and colleagues in the journal *Social Science & Medicine* in January 2014, highlight the importance of considering living arrangements for older adults when it comes to ensuring healthy diets are maintained.

123 **Not everyone** The research from the Faunalytics organization included more than 11,000 Americans with various dietary habits. It can be found at https://faunalytics.org/how-many-former -vegetarians-and-vegans-are-there/.

123 **The healthiest communities** That's according to research from James Enstrom and Lester Breslow at UCLA, who looked at California-dwelling members of the LDS faith between the years of 1980 and 2004 for a study published in *Preventative Medicine* in February 2004.

124 **One of the things** Many young Mormons serve missions, which generally last two years for men and 18 months for women. Men are encouraged to consider mission service at the age of eighteen and women at the age of nineteen.

126 **The "DEAR" approach** Describe, Explain, Ask, and Request. The acronym for this, conveniently enough, is DEAR. And this is what we should do for the ones who are *dear* to us when we are making substantial lifestyle changes.

CHAPTER 4: BE IN MOTION

131 **Studies show the average** An editorial in the *British Journal of Sports Medicine* in February 2009 by Neville Owen and colleagues noted that while research and recommendations have focused on increasing the time that we spend exercising, it's becoming increasingly clear that we need to focus on reducing the amount of time that we spend sitting during our 15.5 "non-exercise" waking hours.

132 **And, quite tragically** I am a big fan of the Designed to Move organization, a collaborative movement launched by the American College of Sports Medicine, the International Council of Sport Science and Physical Education, and Nike, and supported by some 70 other organizations to address the epidemic levels of physical inactivity in the modern world. You can learn more at designedtomove.org. The "movement movement" comes in response to studies like one published by Timothy L. Church and coworkers in *PLOS ONE* in 2011, which concluded that occupation-related energy expenditure has decreased by more than 100 calories over the past few decades, and surmised that this reduction in energy expenditure may account for the increase in US body weights during that same period of time. I don't think that's all of it—but these findings are most certainly a big part of the equation.

133 **By simply comparing** The estimate, published by Mary Shaw and team in the January 2000 edition of *BMJ*, relied on averages and assumed the number of cigarettes smoked throughout a lifetime is constant, among other problems, but researchers noted that it presented the high cost of smoking in a way that everyone can understand.

133 **Far less attention** The study, published by J. Lennert Veerman and colleagues in the BMJ's *British Journal of Sports Medicine* in October 2012, concluded that TV viewing time may be associated with a loss of life that is comparable to other major chronic disease risk factors such as physical inactivity and obesity.

134 **Yes, according to** The estimates roughly correlate to measurements of telomere length, published by Lynn Cherkas and team in the *Archives of Internal Medicine* in January 2008, indicating that spending too much time sitting could cost us ten years of life.

134 **The overwhelming** When Daniela Schmid and Michael Leitzmann did a meta-analysis of 43 studies inclusive of more than 4 million individuals, it was found that the increased likelihood of these
types of cancer among constant sitters could not be undone by exercising. Interestingly, however, lots of sitting was not established to
be related to many other cancers, including breast, stomach, renal
cell, and non-Hodgkin lymphoma. The study was published in the
Journal of the National Cancer Institute in June 2014.

136 **That's what Dr. Loretta DiPietro** This study, which was published
in *Diabetes Care* in October 2013, came as exciting news to me. Many
of my patients walk to lunch already. By simply ensuring they leave
enough time after their meal for a few extra brisk turns around the
block, they are obtaining a tremendous health benefit.

136 **By contrast** The CDC's Adult Participation in Aerobic and Muscle-
Strengthening Physical Activities, which can be found at http://
www.cdc.gov/mmwr/preview/mmwrhtml/mm6217a2.htm,
breaks down the numbers by age, gender, race, education level,
body mass, region, and state—and the results are quite dismal. One
of the best ways to address these issues—and bring ourselves more
in line with a Longevity Village lifestyle—is to have a true assessment of the sorts of exercise we are getting. In a study of more than
6,300 people from across the United States, Richard Troiano and
colleagues demonstrated that actual adherence to physical activity recommendations is substantially lower than what people self-
report. The study was published in *Medicine and Science in Sports and
Exercise* in January 2008.

136 **Indeed, almost all** My patients are certainly not alone. One study,
published by Steven W. Lichtman and coworkers in the *New England
Journal of Medicine* in December 1992, found that obese individuals
overestimated their exercise by about 51 percent and underesti-
mated their caloric intake by 47 percent.

137 **In one of many** The study, which was published by Sindre Mikal
Dyrstad and coworkers in January 2014 in the journal *Medicine and
Science in Sports and Exercise*, lasted seven days and included more
than 1750 participants between the ages of nineteen and eighty-four.

138 **What they learned** The study, published by Duck-chul Lee and
colleagues in the August 5, 2014, issue of the *Journal of the American*

College of Cardiology, concluded that running even just for five minutes at a slow speed conferred significant mortality protection.

140 **My choice of** There was an important caveat offered by Levine and co-author Jennifer Miller in their paper, which was published in May of 2007 in the *British Journal of Sports Medicine*: The consumption of food can't change as a result of increased exercise. The reality is that people do often change their eating habits when they are exercising.

140 **Some people worry** The effect, described by Éadaoin W. Griffin and team in an article in *Physiology and Behavior* in October 2011, showed increased cognitive function specifically in young men; whether the effect is the exact same for other age groups and in women wasn't studied, but my experience and that of many of my patients suggests that everyone can hyperactivate their hippocampal function with aerobic exercise.

140 **Use part of your** A study by Sachin Shah and colleagues in a Canadian hospital demonstrated that taking the stairs rather than the elevator saved about 15 minutes each workday. The study was published in the *Canadian Medical Association Journal* in December 2011.

141 **When it comes to this goal** Fitbit has published its state-by-state breakdown of steps taken at www.fitbit.com/weathermap.

143 **I thought a lot** Swarthmore College professor Barry Schwartz has written a tremendous book about these problems, *The Paradox of Choice*. In it he argues that too many choices lead to unrealistically high expectations, self blame, decision-making paralysis, anxiety, and perpetual stress.

CHAPTER 5: FIND YOUR RHYTHM

149 **About one in four** A quarter of Americans will ultimately experience this heart rhythm disorder according to research by Donald M. Lloyd-Jones and colleagues, which was published in the August 31, 2014, issue of *Circulation*.

In the August 2011 edition of the *Journal of Cardiovascular Electrophysiology*, my partner, T. Jared Bunch, was the lead author and I was the senior author of a report in which we noted that stroke rates were greatly reduced when patients are treated with catheter ablation.

150 **Even in lieu of stroke** Over a five-year period, more than a quarter of the patients we followed for a study published in the April 2010 edition of *HeartRhythm* developed atrial fibrillation and about 4 percent developed dementia. Once again T. Jared Bunch was the lead author and I was the senior author of this publication.

150 **When I started** Medication can be life-saving, but it's hard to adequately describe all of the potential unintended consequences of pharmaceuticals. My colleagues and I, with Victoria Jacobs as the lead author, sought to shine some light on this subject in the December 2014 edition of *HeartRhythm*.

150 **For instance, even though we use these medications** Our recent study of 10,537 patients in the July 2016 issue of the *Journal of the American Heart Association* showed that if the blood thinner, warfarin or Coumadin, wasn't dosed perfectly, the risks of Alzheimer's disease and other forms of dementia were significantly increased.

150 **Today I approach** We looked at those who had been treated with a catheter ablation procedure for an article published in the *Journal of Cardiovascular Electrophysiology* in April 2015, but no matter what means you use to help bring your heart back into rhythm, if you do succeed in doing so, the impact on your health is likely to be positive. My partner, T. Jared Bunch, was the lead author on this study, and I was the senior author for our group.

150 **The overwhelming majority** Long-term sustained weight loss has been shown to be an exceptionally effective way to reduce atrial fibrillation, according to a paper published by Rajeev K. Pathak and colleagues, in the *American Journal of the College of Cardiology* on May 26, 2015.

151 **There are about 50,000 centenarians** Atrial fibrillation is primarily a disease originating from our Western lifestyle. As such, my friend, Sumeet S. Chugh, reported in the February 25, 2014, issue of *Circulation* on how the United States has been hit the hardest by the atrial fibrillation epidemic. Centenarian atrial fibrillation studies have shown similar findings with the United States being hardest hit followed by Europe and the Asians being relatively protected from this disease.

The prevalence of atrial fibrillation in US centenarians was reported to be 27 percent according to an abstract by Saurabh Thakar and colleagues in the September 2012 issue of *Hypertension*. The prevalence of atrial fibrillation was reported to be 16 percent in Danish centenarians according to a study by Karen Andersen-Ranberg and coworkers in the March 2013 issue of *Age Aging*. For Bama, China, one study by Ji-Ying Liang and team in the February 2010 issue of the *Chinese Journal of New Clinical Medicine* reported a prevalence of atrial fibrillation in 3 percent of 120 Bama residents with an average age of 94.5 years old.

152 **Recently, my colleague** I've collaborated on many research efforts with Dr. Bunch. In this one, which was published in the June 2015 edition of the *Journal of Cardiovascular Electrophysiology*, we were looking specifically at the remodeling of the left atrium in patients with atrial fibrillation.

152 **To understand how bad** This evidence, reported by Amneet Sandhu and colleagues in the journal *Open Heart* in March 2014, might be a good reason to end the frivolity of daylight saving time in the modern world. The state of Arizona has already done so; it does not spring forward or fall back in accordance with daylight saving.

153 **A lack of sleep** In a study published by Carla S. Möller-Levet and co-workers in the March 2013 issue of the proceedings of the National Academy of Sciences of the United States of America, researchers showed that people who got just a few hours less sleep each night for a week had an up- or down-regulated alteration in 711 genes. In one specific case, Christian Benedict and team showed sleep deprivation raised neuron-specific enolase, which is produced by small-cell carcinomas, by 20 percent, according to a study published in the journal *Sleep* in January 2014.

153 **And we're doing this** The CDC calls insufficient sleep "a public health epidemic"—and it truly is. You can read more about this at www.cdc.gov/features/dssleep/.

153 **Doctors belong to a culture** A study published by Mirjam Ekstedt and colleagues in the *Journal of Occupational Health Psychology* in April 2012, reveals that preoccupation with thoughts of work during leisure time and high work demands are also high risk factors for burnout, but the biggest predictor of all was lack of sleep.

155 **Before the trip** I love the simplicity of this study, which was

published by Kenneth P. Wright Jr. and coworkers in *Current Biology* in August 2013. All it took was some very basic science and some folks willing to go for a camping trip.

156 **The amount of time** A lot of people aren't even aware of the need to get a good helping of vitamin D each day, which is one of the reasons that vitamin D deficiency is a "serious medical condition that significantly affects the health and well-being of older adults" but which has lacked much attention, according to an analysis published by Megham Meehan and Sue Penckofer in the *Journal of Aging and Gerontology* in December 2014.

158 **Were Naomi and Jermaine** The UCLA Sleep Disorders Clinic, which can be found at www.sleepcenter.ucla.edu, is a wonderful resource for recommendations on how to deal with non-traditional work schedules.

158 **Even small increases** The nutrition educator and writer Mark Nathaniel Mead offered a comprehensive list of scientifically established benefits of sunlight in an article in *Environmental Health Perspectives* in April 2008.

159 **It isn't just sleep** In a study by Ahmed El-Sohemy and colleagues, caffeine slow metabolizers have up to a 64 percent increased risk of a heart attack based on their study published in the October 2, 2007, issue of *Genes and Nutrition*.

159 **Because high caffeine levels** Many people relate intoxication with tough mornings—the hangover effect—but it's important to note that, for many people, just a single drink can impact sleep, as related by Julia K. M. Chan and colleagues in the October 2013 edition of *Alcoholism: Clinical & Experimental Research*.

159 **The people of Longevity Village** I'm quite disinclined to recommend alcohol as part of a healthy lifestyle, but the group of centenarians who have consumed alcohol throughout their lives is large enough that it is clear that moderate drinking does not necessarily have a negative impact on long-term health. And, of course, the group of elders who didn't consume alcohol, according to a study by Zhenyu Xiao and colleagues published in the *Chinese Journal of Population Science* in 1996, was twice as large.

162 **He also set** Irregular meal frequency may lead to weight gain due to a lowered thermic effect of food, according to a study published

by Hamid Farshchi and coworkers in the *International Journal of Obesity and Related Metabolic Disorders* in May 2004.

162 **When we discussed** A study published by Jing Zhang and team in the *Journal of Neuroscience* in March 2014, showed that disrupted sleep is correlated to disturbing levels of degeneration to the locus coeruleus neurons, which are responsible for helping us maintain vigilance and cognitive alertness.

164 **Somehow we've come** A study published by Kep Kee Loh and Ryota Kanai in *PLOS ONE* in 2014 revealed a significant relationship between media multitasking and brain structure variations. Subjects in the study who reported spending more time juggling more than one media device had smaller gray matter density in their anterior cingulate cortex.

164 **We *are* amazing creatures** Dave Crenshaw takes a prosecutorial approach to dismantling the notion that we can learn to be good multitaskers in *The Myth of Multitasking: How "Doing It All" Gets Nothing Done*. I highly recommend this book to patients who are suffering under the weight of trying to do too much at once.

164 **This might be best exemplified** I loved *M*A*S*H*, and have a particularly soft spot in my heart for the character of Charles Winchester, a masterful surgeon who was played to perfection, complete with a perfect Boston Brahmin accent, by the Juilliard-trained actor David Ogden Stiers.

164 **Do you think** It's clear that mutlitasking doesn't just impact our ability to do things well, it impacts our ability to self-assess how we do things, as demonstrated in a study published by Eyal Ophir and colleagues in the September 2009 edition of the Proceedings of the National Academy of Sciences of the United States of America.

165 **Multitasking doesn't just** While a study, published by Loh and Kanai in the journal *PLOS ONE* in September 2014, did not establish whether decreases in ACC gray matter were a result of multitasking or that people were more likely to attempt to multitask if they had less gray matter, the correlation should be cause for great concern.

165 **And since we're talking** David L. Strayer's work, including a study he published with colleagues in the summer 2006 edition of *Human Factors*, should have prompted legislators across the country to take notice of the epidemic in distracted driving. While Strayer can

certainly take pride in the impact of his work (indeed, it has undoubtedly saved lives) lawmakers should be ashamed at how they
have willfully ignored these findings and responded with such a
lack of haste.

CHAPTER 6: MAKE THE MOST OF YOUR ENVIRONMENT

169 **Sometime around her 105th birthday** It's probably important to
put the tourism in Bapan into context. It often seems like the village
is being overwhelmed by Chinese tourists, but a visit to Longevity Village is still a very rare treat. There are 1.35 billion people in
China, and each year more of them are traveling domestically. In
2014, China's 26,000 travel agencies served more than 131 million domestic tourists. In a very busy day in Bapan, there might be an extra
100 tourists in the village. Even if that happened every day—and it
certainly doesn't—it would still represent just one person for every
3,500 domestic tourists in China.

171 **The cleaning products** The Environmental Working Group's
Cleaners Database is a great place to go for information about the
relative safety of various products. http://www.ewg.org/cleaners
/hallofshame/.

174 **I think it would be fun** David P. Strachan was one of the first researchers to promote the Hygiene Hypothesis in an article he published in the *British Medical Journal* on November 18, 1989. In this
landmark article, he reported that less early childhood infections
was associated with more asthma and hay fever.

175 **Our twentieth-century obsession** According to a report by Greg-
ory L. Armstrong and colleagues in the *Journal of the American Medical Association* in 1999, infectious disease mortality declined during
the first eight decades of the twentieth century from 797 deaths per
100,000 in 1900 to 36 deaths per 100,000 in 1980.

175 **Many modern in-home dishwashers** It could be easy to read too
much into this study, which was based on 1,029 children aged seven
to eight years from Kiruna, in the north of Sweden, and Mölndal,
in the Gothenburg area on the southwest coast of Sweden. It was
published by Bill Hesselmar and coworkers in the journal *Pediatrics*

in March 2015. It was, however, a fascinating finding and one that has prompted many conversations about what cleanliness really means.

176 **Bill Hesselmar** In a study published in *Pediatrics* in June 2013, researchers speculated that the positive health effects seen in children whose parents sucked their pacifiers, instead of washing them, might be due to immune stimulation by microbes transferred to the infant via the parent's saliva. Just as fascinating, a study published in the *Proceedings of the National Academy of Sciences of the United States of America* in January 2014, found a "distinct gastrointestinal microbiome composition" in mice exposed to dust from homes where there was a dog present.

177 **The Chinese government** A study published by Ranjini M. Krishnan and team in the *Journal of the American College of Cardiology* in November 2012 evaluated the association of long- and short-term air pollutant exposures with dilation of the brachial artery. The results discredited any notion that short-term air pollution was a short-term health concern.

177 **This is also a tremendous problem** The city rankings for the 2015 State of the Air report are available at www.stateoftheair.org/2015/city-rankings/most-polluted-cities.html.

178 **Qui Yi's experience** In one study showing a correlation between pollution and poor sleep, Antonella Zanobetti and colleagues used data from the Sleep Heart Health Study, which involved thousands of subjects (all of them over the age of thirty-nine) to examine whether particulate air matter was associated with sleep-disordered breathing. It was. The study was published in the *American Journal of Respiratory and Critical Care Medicine* in September 2010.

179 **One significant point** Advances in mapping technology and sophistication have given us incredible information about public health. In one study, published by Takashi Yamashita and Suzanne Kunkel in *Social Science & Medicine* in 2010, a geographic information system approach was used to integrate, visualize, and analyze data from hospitals in all of Ohio's 88 counties—an effort that would have been all but impossible before computer data analysis.

180 **One patient** At this point, the link between air pollution with heart

attack and stroke is undeniable. In one study, published by Sara D. Adar and colleagues in *PLOS Medicine* in April 2013, it was demonstrated that higher levels of fine particular air pollution were associated with faster progression of intima-medial thickness.

180 **About 40 percent** There may be some exceptions to the escape-to-the-beach-for-clean-air approach, particularly in industrial beach areas.

181 **In fact, this might be** It simply doesn't make a lot of sense to tell people to go inside when the air is bad; unless we're making changes that positively impact the air inside of our homes, going inside isn't helpful.

182 **Houseplants can be** A reasonably legible photocopied version of the report can be viewed on NASA's document archive website: http://ntrs.nasa.gov/archive/.

182 **Have you ever** VOCs can also cause loss of concentration, nausea and eye, nose, and throat irritation, according to the federal Environmental Protection Agency. My cowriter's sister, eco-interior designer Kelly LaPlante, shared the "this smells new" rule with us; it's one I now share with many of my patients—and it goes for anything we buy. You can see more ways in which your home can be both beautiful and healthy at www.kellylaplante.com/.

182 **The same goes for furniture** A study published by Heather B. Patisaul and coworkers in the *Journal of Biochemistry and Molecular Toxicology* in February 2013, implicated Firemaster 550 as an endocrine disruptor, with effects including those I've already named plus increased serum thyroxine levels, reduced hepatic carboxylesterease activity, and altered exploratory behaviors in offspring.

183 **I also advise** When physicist, data scientist, and avid runner Lyndie Chiou wanted to know why she sometimes found herself coughing like a chain smoker after going for an evening run, she contacted her local air quality agencies in the San Francisco Bay Area, sorted through the data, and came upon some interesting conclusions. Her website is highly recommended: www.researchpipeline.com/.

184 **In most American cities** The *New York Times*, in a fascinating and troubling piece of public service journalism, measured noise levels at 37 locations across the city. In an article published on July 19, 2012, journalist Cara Buckley described "levels that experts said bordered on dangerous" at a third of the places tested.

185 **The simple trappings** One of the pioneers in this field is someone with plenty of life experience in sensory excess, a well-known actress, model, and "night-lifer" from Tel Aviv named Nilli Lavie. Her work with James Macdonald, which was published in *Attention, Perception, and Psychophysics* in May 2011, broke exciting new ground on inattention, adding deafness to blindness in a list of senses that can be selectively shut down by the brain when humans are focused on something else.

185 **When we think of the impact** The relationship of hearing impairment to dementia and cognitive decline was first reported by Richard F. Uhlmann and colleagues in the April 7, 1989, issue of the *Journal of the American Medical Association.*

185 **Researchers keep finding** The authors of a study published by Jaana I. Halonen and team in the *European Heart Journal* on October 14, 2015, concluded that long-term exposure to road traffic noise was associated with small increased risks of all-cause mortality and cardiovascular mortality, particularly for stroke in the elderly.

185 **Determining the decibel level** While not definitive—the authors did note that alternative explanations such as ecological contamination should be considered—a study published by Anna Hansell and colleagues in the *British Medical Journal* on October 8, 2013, should be concerning for anyone who lives in an area with frequently high levels of noise. In the study, researchers found that aircraft noise was associated with an increased risk for stroke and cardiovascular disease.

186 **The evidence of a connection** There appears to be a significantly increased risk of heart problems in subjects who live in areas with outdoor noise levels greater than 65 decibels, according to an epidemiological review by Wolfgang Babisch published in *Noise and Health* in 2000. That's a far lower sound threshold than exists on many city streets.

186 **Whether or not** More panes, thicker glass, and even glazing may significantly determine the internal noise of your home, according to a study by Bin Guo and colleagues published in *Beijing Da Xue Xue Bao* in the *Journal of Peking University Health Sciences,* June 2015.

188 **Because the vast majority** The Colorado River, which serves as a major source of drinking water for Los Angeles, San Diego,

Phoenix, and many other cities and towns, was found to be contaminated by a type of rocket fuel known as perchlorate, among other chemicals, from an industrial site in Henderson, Nevada, according to a 2003 report from the Natural Resource Defense Council.

189 **My colleagues and I** My colleagues and I have published extensively over the years on the relationship of cardiovascular disease to dementia, especially that of atrial fibrillation. Our landmark study on this topic, for which T. Jared Bunch was the lead author and I was the senior author, was titled "Atrial fibrillation is independently associated with senile, vascular, and Alzheimer's dementia" and was published in the April 2009 issue of *HeartRhythm*.

191 **Three-quarters of middle-class residents** In a creative and fascinating study published in the January 2010 issue of *Personality and Social Psychology Bulletin*, researchers Darby Saxbe and Rena Repetti were taken on tours of Angelinos' houses by female residents, whose reactions to showing a stranger their homes were recorded on video. The women used words like "mess," "not fun," and "very chaotic" to describe their homes. Using saliva samples, the researchers measured levels of diurnal cortisol, and found a link between how people talk about their homes and their levels of stress.

191 **Consider the desktop** Any time there is an abundance of visual stimuli, there is a competition for brain attention, according to a study by Stephanie McMains and Sabine Kastner published in the *Journal of Neuroscience* on January 12, 2011.

192 **I'm certainly not** When cutting back on clutter, look for thrift stores that support a non-profit cause you support. It's a great way to give.

194 **Having the right** Certain cancer-fighting drugs, it has been found, induce the translocation of selected species of bacteria into secondary lymphoid organs. There, the bacteria stimulate the generation of pathogenic T-cells, promoting an increased immune response, according to a study by Sophie Viaud and colleagues published in *Science* on November 22, 2013.

194 **There's also compelling** A study of centenarians and so-called "younger elderly" (those between the ages of eighty-five and ninety-nine) of Bama, China, published by Fang Wang and colleagues in the *Journal of Microbiology and Biotechnology* in April 2015,

showed that a high-fiber diet was an essential element to creating a healthy gut biome which, in turn, promotes a more healthy life. The human gut microbiome is often shared among family members as a result of shared environments and diets, though each person's gut flora varies considerably, according to a paper by Peter J. Turnbaugh and coworkers published in the journal *Nature* on January 22, 2009.

195 **As a cardiologist** One study, published by Wayne A. Ray and team in the May 17, 2012, issue of the *New England Journal of Medicine*, noted an increase in cardiovascular deaths over a five-day period of azithromycin therapy among patients with a high risk of cardiovascular disease. Another study, published by Ben Boursi and colleagues in the June 2015 edition of the *European Journal of Endocrinology*, demonstrated that exposure to certain antibiotic groups was linked to an increased diabetes risk; that study was enormous, with 208,002 diabetic cases and 815,576 matched controls.

CHAPTER 7: PROCEED WITH PURPOSE

204 **It has been** China's "one child policy" has also now been reformed so that many families can now have two children, particularly those in which both parents are single children themselves. Many ethnic minorities have also long had the right under Chinese law to have more than one child if the first is a girl.

207 **Holocaust survivor** Victor Frankl's book, *Man's Search for Meaning*, was originally published in German as *Trotzdem Ja Zum Leben Sagen: Ein Psychologe Erlebt das Konzentrationslager*, or, roughly translated: *Nevertheless, Say Yes to Life: A Psychologist Experiences the Concentration Camp*. It is fascinating reading.

207 **The challenges faced** A study published by Patricia A. Boyle and colleagues in the *Archives of General Psychiatry* in May 2012, included 246 older adults from the Rush Memory and Aging Project. It looked at not only qualitative measures of purpose, but also included postmortem examinations of patient's brains to determine the extent to which there were pathologic changes indicative of Alzheimer's disease. These sorts of findings appear to be universal; researchers writing for *Psychosomatic Medicine* in June 2009, noted that while myriad life variables impacted people's sense of purpose differently

along racial, ethnic, gender, and educational lines, the overall impact on life mortality remained the same.

208 **As is often the case** It's important to remember that depression is a treatable disorder, and when we address a person's depression, we can lower their risk of another myocardial infarction event, according to a paper by Redford B. Williams published in the journal *Circulation* on June 28, 2011.

210 **Stress, after all** Chinese medical students, perhaps some of the most stressed-out people in the world, were studied to better understand what evidence-based intervention strategies might best enhance their resilience in order to promote better life satisfaction. The study was published by Meng Shi and colleagues in the February 2013 edition of *BioMed Central Medical Education*.

210 **Having a strong sense** Susan A. Everson and coworkers writing for the journal *Arteriosclerosis, Thrombosis, and Vascular Biology* in August 17, 1997, noted that hopelessness contributes to the progression of carotid atherosclerosis and that chronic hopelessness can be especially detrimental.

211 **There was a group** Mark Murivan's study, which was published in the *Journal of Research in Personality* in August 2008, is a classic example of how we don't need huge budgets or expensive laboratories to engage in meaningful research. All Murivan needed was some hungry undergrads, an empty room, and some chocolate chip cookies.

214 **Feeling needed** A meta-analysis of the impact of volunteerism on health by Caroline E. Jenkinson and team was published in *BMC Public Health* in August 2013. It revealed some fascinating things about the positive impact of volunteering, including longer lives among volunteers.

215 **One of the reasons** Epidemiologist Michael Marmot, who wrote about the influence of income on health in the March 2002 edition of *Health Affairs*, concluded that income inequality was such a problem for public health that governments needed to take a stronger approach to addressing it.

216 **When people think** The research on the connection between future-self-visualization and financial planning is fascinating, and represents just the tip of the iceberg when it comes to ways in which technology can help us better prepare for what's to come. The study

was published in the *Journal of Marketing Research* in 2011 by Hal E. Hershfield and colleagues.

217 **There's tremendously good evidence** A study published by Robert Hammerman-Rozenberg and colleagues in *Aging Clinical and Experimental Research* in 2005, demonstrated that both men and women who work in their senior years have better perceived health and greater independence.

CHAPTER 8: LONG LIVE THE VILLAGE

230 **A long-running study out of Harvard University** For example, the Physicians Health Study that Laurel B. Yates and coworkers published in the *Archives of Internal Medicine* in February 2008 noted that things like smoking, weight gain and loss, and blood pressure, were "modifiable healthy behaviors." That's a good way of emphasizing that all of the things that seemed to result in long and healthy lives were quite under the control of the study participants.

BIBLIOGRAPHY

(* Research to which Dr. John Day contributed)

Abbott, A. (2014) Sugar substitutes linked to obesity. *Nature*.

Adar, S., et. al. (2013) Fine particulate air pollution and the progression of carotid intima-medial thickness: A prospective cohort study from the multi-ethnic study of atherosclerosis and air pollution. *PLOS Medicine*.

Andersen-Ranberg K., et. al. (2013) Cardiovascular diseases are largely underreported in Danish centenarians. *Age Aging*.

Armstrong, G., Conn, L., and Pinner, R. (1999) Trends in infectious disease mortality in the United States during the 20th century. *Journal of the American Medical Association*.

Babisch, W. (2000) Traffic noise and cardiovascular disease: Epidemiological review and synthesis. *Noise Health*.

Barrett, D. (2007) Maximizing the nutritional value of fruits and vegetables. *Food Technology*.

Benedict, C., et. al. (2014) Acute sleep deprivation increases serum levels of neuron-specific enolase (NSE) and S100 calcium binding protein B (S-100B) in healthy young men. *Sleep*.

Bibbins-Domingo, K. (2014) Sodium intake in populations: Assessment of evidence. *Journal of the American Medical Association: Internal Medicine*.

Boschmann, M., et. al. (2003) Water-induced thermogenesis. *Journal of Clinical Endocrinology and Metabolism*.

Boursi, B., et. al. (2015) The effect of past antibiotic exposure on diabetes risk. *European Journal of Endocrinology*.

Boyle, P., et. al. (2009) Purpose in life is associated with mortality among community-dwelling older persons. *Psychosomatic Medicine.*

Boyle, P., et. al. (2012) Effect of purpose in life on the relation between Alzheimer disease pathologic changes on cognitive function in advanced age. *Archives of General Psychiatry.*

Buckley, C. (July 7, 2012) In New York City, indoor noise goes unabated. *The New York Times.*

Buettner, D. (2010) *The Blue Zones: Lessons for Living Longer from the People Who've Lived the Longest.* National Geographic.

* Bunch T., et. al. (2009) Atrial fibrillation is independently associated with senile, vascular, and Alzheimer's dementia. *Heart Rhythm.*

* Bunch, T., et. al. (2010) Patients treated with catheter ablation for atrial fibrillation have long-term rates of death, stroke, and dementia similar to patients without atrial fibrillation. *Heart Rhythm.*

* Bunch, T., et. al. (2011) Patients treated with catheter ablation for atrial fibrillation have long-term rates of death, stroke, and dementia similar to patients without atrial fibrillation. *Journal of Cardiovascular Electrophysiology.*

* Bunch, T. and Day, J. (2015) Adverse remodeling of the left atrium in patients with atrial fibrillation: When is the tipping point in which structural changes become permanent? *Journal of Cardiovascular Electrophysiology.*

* Bunch, T., et. al. (2015) Five-year outcomes of catheter ablation in patients with atrial fibrillation and left ventricular systolic dysfunction. *Journal of Cardiovascular Electrophysiology.*

* Bunch, T., et al. (2016) Atrial fibrillation patients treated with long-term warfarin anticoagulation have higher rates of all dementia types compared with patients receiving long-term warfarin for other indications. *Journal of the American Heart Association.*

Campbell, T. and Campbell, T. (2006) *The China Study: Startling Implications for Diet, Weight Loss and Long-term Health.* BenBella Books.

Campbell T., Parpia, B., and Chen J. (1998) Diet, lifestyle, and the etiology of coronary artery disease: the Cornell China study. *American Journal of Cardiology.*

Centers for Disease Control and Prevention (2015) *Insufficient Sleep Is a Public Health Problem.* www.cdc.gov.

Chan, J., et. al. (2013) The acute effects of alcohol on sleep architecture in late adolescence. *Alcoholism, Clinical and Experimental Research.*

Chen, N-Y., et. al. (2014) Enrichment of MTHFR 677 T in a Chinese long-lived cohort and its association with lipid modulation. *Lipids in Health and Disease.*

Cherkas L., et. al. (2008) The association between physical activity in leisure time and leukocyte telomere length. *Archives of Internal Medicine.*

Chiou, L. (2014) *The Best Time of Day for Your Lungs to Exercise.* www .researchpipeline.com.

Chiuve, S., et. al. (2011) Plasma and dietary magnesium and risk of sudden cardiac death in women. *The American Journal of Clinical Nutrition.*

Chugh, S., et. al. (2014) Worldwide epidemiology of atrial fibrillation: a Global Burden of Disease 2010 Study. *Circulation.*

Church, T., et. al. (2011) Trends over 5 decades in US occupation-related physical activity and their associations with obesity. *PLOS ONE.*

Collier, R. (2013) Intermittent fasting: the science of going without. *Canadian Medical Association Journal.*

Conklin, A., et. al. (2014) Social relationships and healthful dietary behaviour: Evidence from over-50s in the EPIC cohort, UK. *Social Science & Medicine.*

Crenshaw, D. (2008) *The Myth of Multitasking: How "Doing It All" Gets Nothing Done.* Jossey-Bass.

The Dalai Lama (1998) *The Four Noble Truths.* Thorsons.

Darmadi-Blackberry I., et. al. (2004) Legumes: the most important dietary predictor of survival in older people of different ethnicities. *Asia Pacific Journal of Clinical Nutrition.*

Davis, D., Epp, M., and Riordan, H. (2003) Changes in USDA food composition data for 43 garden crops, 1950 to 1999. *Journal of the American College of Nutrition.*

Davy, B., et. al. (2008) Water consumption reduces energy intake at a breakfast meal in obese older adults. *Journal of the American Dietetic Association.*

Dhahbi, J., et. al. (2004) Temporal linkage between the phenotypic and genomic responses to caloric restriction. *Proceedings of the National Academy of Sciences of the United States of America.*

Dhanwal, D., et. al. (2011) Epidemiology of hip fracture: Worldwide geographic variation. *Indian Journal of Orthopaedics.*

DiPietro, L., et. al. (2013) Three 15-min bouts of moderate postmeal walking significantly improves 24-h glycemic control in older people at risk for impaired glucose tolerance. *Diabetes Care.*

Domínguez-Rodrigo M., et. al. (2012) Earliest porotic hyperostosis on a 1.5-million-year-old hominin, olduvai gorge, Tanzania. *PLOS ONE.*

Dyrstad, S., et. al. (2014) Comparison of self-reported versus accelerometer-measured physical activity. *Medicine and Science in Sports and Exercise.*

Eichstaedt, J., et. al. (2015) Psychological language on Twitter predicts county-level heart disease mortality. *Psychological Science.*

Ekstedt, M., et. al. (2006) Disturbed sleep and fatigue in occupational burnout. *Scandinavian Journal of Work, Environment, and Health.*

El-Sohemy, A., et. al. (2007) Coffee, CYP1A2 genotype and risk of myocardial infarction. *Genes and Nutrition.*

Enstrom, J. and Breslow, L. (2008) Lifestyle and reduced mortality among active California Mormons, 1980–2004. *Preventative Medicine.*

Epel, E., et. al. (2004) Accelerated telomere shortening in response to life stress. *Proceedings of the National Academy of Sciences of the United States of America.*

Epton T., et. al. (2015) The impact of self-affirmation on health-behavior change: a meta-analysis. *Health Psychology.*

Estruch, R., et. al. (2013) Primary prevention of cardiovascular disease with a Mediterranean diet. *New England Journal of Medicine.*

Everson S. (1997) Hopelessness and 4-year progression of carotid atherosclerosis. The Kuopio ischemic heart disease risk factor study. *Arteriosclerosis, Thrombosis, and Vascular Biology.*

Fagan R. and Fagan J. (2009) Play behaviour and multi-year juvenile survival in free-ranging brown bears, Ursus arctos. *Evolutionary Ecology Research.*

Farshchi, H., Taylor, M., and Macdonald, I. (2004) Decreased thermic effect of food after an irregular compared with a regular meal pattern in healthy lean women. *International Journal of Obesity and Related Metabolic Disorders.*

Fei X., Wu, J., Kong, Z., and Christakos, G. (2015) Urban-rural disparity of breast cancer and socioeconomic risk factors in China. *PLOS ONE.*

Fowler, G. (Feb. 17, 2016) Texting while walking isn't funny anymore. *The Wall Street Journal.*

Frankl, V. (2006) *Man's Search for Meaning.* Beacon Press.

Fujimura, K. E., et. al. (2014) House dust exposure mediates gut microbiome lactobacillus enrichment and airway immune defense against allergens and virus infection. *Proceedings of the National Academy of Sciences of the United States of America.*

Gibson, L. and Benson, G. (2002) *Origin, History, and Uses of Corn (Zea mays).* Iowa State University Department of Agronomy.

Griffin É., et. al. (2011) Aerobic exercise improves hippocampal function and increases BDNF in the serum of young adult males. *Physiology and Behavior.*

Guo, B., Huang, J. and Guo, X. (2015) Preventive effects of sound insulation windows on the indoor noise levels in a street residential building in Beijing. *Da Xue Xue Bao, the Journal of Peking University Health Sciences.*

Halonen, J., et. al. (2015) Road traffic noise is associated with increased cardiovascular morbidity and mortality and all-cause mortality in London. *European Heart Journal.*

Hammerman-Rozenberg, R., et. al. (2005) Working late: the impact of work after 70 on longevity, health and function. *Aging Clinical and Experimental Research.*

Hansell A., et. al. (2013) Aircraft noise and cardiovascular disease near Heathrow airport in London: small area study. *British Medical Journal.*

Hershfield, H., et. al. (2011) Increasing saving behavior through age-progressed renderings of the future self. *Journal of Marketing Research.*

Herskind, A., et. al. (1996) The heritability of human longevity: a population-based study of 2872 Danish twin pairs born 1870–1900. *Human Genetics.*

Hertenstein, M., et. al. (2009) Smile intensity in photographs predicts divorce later in life. *Motivation and Emotion.*

Hesselmar, B., et. al. (2013) Pacifier cleaning practices and risk of allergy development. *Pediatrics.*

Hesselmar, B., Hicke-Roberts, A., and Wennergren G. (2015) Allergy in children in hand versus machine dishwashing. *Pediatrics.*

Holt, S., Miller, J., Petocz, P., and Farmakalidis, E. (1995) A satiety index of common foods. *European Journal of Clinical Nutrition.*

Holt-Lunstad, J., et. al. (2003) Social relationships and ambulatory blood pressure: structural and qualitative predictors of cardiovascular function during everyday social interactions. *Health Psychology.*

Horne, B., et. al. (2012) Relation of routine, periodic fasting to risk of diabetes mellitus, and coronary artery disease in patients undergoing coronary angiography. *American Journal of Cardiology.*

Huang, T., et. al. (2015) Consumption of whole grains and cereal fiber and total and cause-specific mortality: prospective analysis of 367,442 individuals. *BMC Medicine.*

* Jacobs, V., et. al. (2014) Time outside of therapeutic range in atrial fibrillation patients is associated with long-term risk of dementia. *Heart Rhythm.*

Jenkinson, C., et. al. (2013) Is volunteering a public health intervention? A systematic review and meta-analysis of the health and survival of volunteers. *BMC Public Health.*

Jung, K., et. al. (2011) Effects of floor level and building type on residential levels of outdoor and indoor polycyclic aromatic hydrocarbons, black carbon, and particulate matter in New York City. *Atmosphere.*

Kasulis, T. (1989) *Zen Action/Zen Person.* University of Hawaii Press.

Keller, A. (2012) Does the perception that stress affects health matter? The association with health and mortality. *Health Psychology.*

Kiecolt-Glaser, J. and Glaser, R. (2010) Psychological stress, telomeres, and telomerase. *Brain, Behavior, and Immunity.*

Koepp G. (2013) Treadmill desks: A 1-year prospective trial. *Obesity.*

Krishnan, R., et. al. (2012). Vascular responses to long- and short-term exposure to fine particulate matter: MESA air (Multi-ethnic study of atherosclerosis and air pollution). *Journal of the American College of Cardiology.*

Lampert, R., et. al. (2014) Triggering of symptomatic atrial fibrillation by negative emotion. *Journal of the American College of Cardiology.*

Lawrence, G. (2013) Dietary fats and health: dietary recommendations in the context of scientific evidence. *Advanced Nutrition.*

Lee, D., et. al. (2014) Leisure-time running reduces all-cause and cardiovascular mortality risk. *Journal of the American College of Cardiology.*

Lee, S. and Schwarz, N. (2011) Wiping the slate clean: Psychological consequences of physical cleansing. *Current Directions in Psychological Science.*

Levine, J. and Miller, J. (2007) The energy expenditure of using a "walk-and-work" desk for office workers with obesity. *British Journal of Sports Medicine.*

Levy B., et. al. (2002) Longevity increased by positive self-perceptions of aging. *Journal of Personality and Social Psychology.*

Liang J., et. al. (2010) Analysis of electrocardiogram in 120 longevity elderly in Bama area. *Chinese Journal of New Clinical Medicine.*

Lichtman, S., et. al. (1992) Discrepancy between self-reported and actual caloric intake and exercise in obese subjects. *New England Journal of Medicine.*

Lloyd-Jones, D., et. al. (2004) Lifetime risk for development of atrial fibrillation: the Framingham Heart Study. *Circulation.*

Loh, K. and Kanai, R. (2014) Higher media multi-tasking activity is associated with smaller gray-matter density in the anterior cingulate cortex. *PLOS ONE.*

Ludy, M., Moore, G., and Mattes, R. (2011) The effects of capsaicin and capsiate on energy balance: critical review and meta-analyses of studies in humans. *Chemical Senses.*

Lunstad J., et. al. (2003) Social relationships and ambulatory blood pressure: structural and qualitative predictors of cardiovascular function during everyday social interactions. *Health Psychology.*

Lunstad J., et. al. (2015) Loneliness and social isolation as risk factors for mortality: a meta-analytic review. *Perspectives on Psychological Science.*

Luu, H., et. al. (2015) Prospective evaluation of the association of nut/peanut consumption with total and cause-specific mortality. *JAMA Internal Medicine.*

Macdonald, J. and Lavie, N. (2011) Visual perceptual load induces inattentional deafness. *Attention, Perception & Psychophysics.*

Marmot, M. (2002) The influence of income on health: views of an epidemiologist. *Health Affairs.*

Martin, P., et. al. (2006) Personality and longevity: findings from the Georgia Centenarian Study. *Age.*

Mattes, R. (2010) Hunger and thirst: Issues in measurement and prediction of eating and drinking. *Physiology & Behavior.*

McGregor, J. (Sept. 2, 2014) The average workweek is now 47 hours. *The Washington Post.*

McMains, S. and Kastner, S. (2011) Interactions of top-down and bottom-up mechanisms in human visual cortex. *The Journal of Neuroscience: The Official Journal of the Society for Neuroscience.*

McPherson, M., Smith-Lovin, L., and Brashears, M. (2006) Social isolation in America: Changes in core discussion networks over two decades. *American Sociological Review.*

Mead, M. (2008) Benefits of sunlight: A bright spot for human health. *Environmental Health Perspectives.*

Meehan, M. and Penckofer, S. (2014) The role of vitamin D in the aging adult. *Journal of Aging and Gerontology.*

Moalem, S. and LaPlante, M. (2014) *Inheritance: How Our Genes Change Our Lives, and Our Lives Change Our Genes.* Grand Central Publishing.

Möller-Levet, C., et. al. (2013) Effects of insufficient sleep on circadian rhythmicity and expression amplitude of the human blood transcriptome. *Proceedings of the National Academy of Sciences of the United States of America.*

Mozaffarian, D. (2011) Changes in diet and lifestyle and long-term weight gain in women and men. *New England Journal of Medicine.*

Mozaffarian, D., et. al. (2013) Plasma phospholipid long-chain omega-3 fatty acids and total and cause-specific mortality in older adults: the cardiovascular health study. *Annals of Internal Medicine.*

Mund M. and Mitte, K. (2012) The costs of repression: a meta-analysis on the relation between repressive coping and somatic diseases. *Health Psychology.*

Muraven, M. (2008) Autonomous self-control is less depleting. *Journal of Research in Personality.*

Natural Resource Defense Council (2003) *Source water protection: Findings and recommendations.*

Nawijn, J., et. al. (2010) Vacationers happier, but most not happier after a holiday. *Applied Research in Quality of Life.*

Nielsen, R., et. al. (2014) A prospective study on time to recovery in 254 injured novice runners. *PLOS ONE.*

Ning-Yuan, C., et. al. (2014) Enrichment of MTHFR 677 T in a Chinese long-lived cohort and its association with lipid modulation. *Lipids in Health and Disease.*

NPD Group (2014) *Consumers are Alone Over Half of Eating Occasions as a*

Result of Changing Lifestyles and More Single-person Households. www
.npd.com.

O'Donovan, A., et. al. (2009). Pessimism correlates with leukocyte
telomere shortness and elevated interleukin-6 in post-menapausal
women. *Brain, Behavior, and Immunity.*

Ophir, E., et. al. (2009) Cognitive control in media multitaskers. *Proceedings of the National Academy of Sciences of the United States of America.*

Otto, M. and Smits, J. (2011) *Exercise for Mood and Anxiety.* Oxford University Press.

Owen, N., Bauman, A., and Brown, W. (2008) Too much sitting: a novel
and important predictor of chronic disease risk? *British Journal of
Sports Medicine.*

Pappas, S. (Oct. 21, 2013) Oreos are as enticing as cocaine, a rat study
finds. But don't worry about withdrawal. *The Washington Post.*

Parretti, H., et. al. (2015) Efficacy of water preloading before main meals
as a strategy for weight loss in primary care patients with obesity.
Obesity.

Paschal, S. (2006) *Does changing cognitions cause health behaviour change?*
Conference of the European Health Psychology Society; Warsaw,
Poland.

Pathak, R., et. al. (2015) Long-term effect of goal-directed weight management in an atrial fibrillation cohort: A long-term follow-up study
(LEGACY). *American Journal of the College of Cardiology.*

Patisaul, H., et. al. (2013) Accumulation and endocrine disrupting effects
of the flame retardant mixture Firemaster® 550 in rats: An exploratory assessment. *Journal of Biochemical Molecular Toxicology.*

Philippot, P., Chapelle, G., and Blairy S. (2002) Respiratory feedback in
the generation of emotion. *Cognition & Emotion.*

Rada, P., Avena, N., and Hoebel, B. (2005) Daily bingeing on sugar repeatedly releases dopamine in the accumbens shell. *Neuroscience.*

Ralston, N. and Raymond, L. (2010) Dietary selenium's protective effects
against methylmercury toxicity. *Toxicology.*

Ray, W., et. al. (2012) Azithromycin and the risk of cardiovascular death.
The New England Journal of Medicine.

Sabaté, J. (1999) Nut consumption, vegetarian diets, ischemic heart disease risk, and all-cause mortality: evidence from epidemiologic studies. *American Journal of Clinical Nutrition.*

Sandhu, A., Seth, M., and Gurm, H. (2014) Daylight savings time and myocardial infarction. *Open Heart.*

Saxbe, D. and Repetti, R. (2010) No place like home: home tours correlate with daily patterns of mood and cortisol. *Personality & Social Psychology.*

Schmid, D. and Leitzmann, M. (2014) Television viewing and time spent sedentary in relation to cancer risk: a meta-analysis. *Journal of the National Cancer Institute.*

Schwartz, B. (2005) *The Paradox of Choice: Why Less Is More.* Harper Perennial.

Sell, A., et al. (2014) The human anger face evolved to enhance cues of strength. *Evolution and Human Behavior.*

Shah, S., et. al. (2011) Elevators or stairs? *Canadian Medical Association Journal.*

Shaw, M., Mitchell, R., and Dorling, D. (2000) Time for a smoke? One cigarette reduces your life by 11 minutes. *British Medical Journal.*

Shi, M., et. al. (2015) The mediating role of resilience in the relationship between stress and life satisfaction among Chinese medical students: a cross-sectional study. *BMC Medical Education.*

Shirom, A., et. al. (2011) Work-based predictors of mortality: a 20-year follow-up of healthy employees. *Health Psychology.*

Shoot, B. (March 24, 2014) Water buffalo wade in the American dairy scene. *Modern Farmer.*

* Smith, M., et. al. (2011) Vitamin D excess is significantly associated with risk of atrial fibrillation (abstract). *Circulation.*

Söderström, M., et. al. (2012) Insufficient sleep predicts clinical burnout. *Journal of Occupational Health.*

Strachan, D. (1989) Hay fever, hygiene, and household size. *British Medical Journal.*

Strayer, D., Drews F., and Crouch D. (2006) A comparison of the cell phone driver and the drunk driver. *Human Factors.*

Suez J., et. al. (2014) Artificial sweeteners induce glucose intolerance by altering the gut microbiota. *Nature.*

Sugiyama, T., et. al. (2014) Is there gluttony in the time of statins? Different time trends of caloric and fat intake between statin-users and non-users among US adults. *Journal of the American Medical Association: Internal Medicine.*

Taormina, G. and Mirisola, M. (2015) Longevity: epigenetic and biomolecular aspects. *Biomolecular Concepts.*

Thakar S., et. al. (2012) Electrocardiographic changes in patients ≥ 100 years of age (abstract). *Hypertension.*

Troiano, R., et. al. (2008) Physical activity in the United States measured by accelerometer. *Medicine and Science in Sports and Exercise.*

Turnbaugh, P., et. al. (2009) A core gut microbiome in obese and lean twins. *Nature.*

Uhlmann R., et. al. (1989) Relationship of hearing impairment to dementia and cognitive dysfunction in older adults. *Journal of the American Medical Association* (JAMA).

Veerman, J., et. al. (2012) Television viewing time and reduced life expectancy: a life table analysis. *British Journal of Sports Medicine.*

Viaud, S., et. al. (2013) The intestinal microbiota modulates the anticancer immune effects of cyclophosphamide. *Science.*

Walleghen E., et. al. (2007) Pre-meal water consumption reduces meal energy intake in older but not younger subjects. *Obesity.*

Wang F., et. al. (2015) Gut microbiota community and its assembly associated with age and diet in Chinese centenarians. *Journal of Microbiology and Biotechnology.*

Wilcox D., et. al. (2006) Genetic determinants of exceptional human longevity: insights from the Okinawa Centenarian Study. *Age.*

Williams R. (2011) Depression after heart attack: why should I be concerned about depression after a heart attack? *Circulation.*

Wilson, R., et. al. (2007) Loneliness and risk of Alzheimer disease. *Archives of General Psychiatry.*

Wing, R. and Jeffery, R. (1999) Benefits of recruiting participants with friends and increasing social support for weight loss and maintenance. *Journal of Consulting and Clinical Psychology.*

Wolverton, B., Johnson, A., and Bounds, K. (1989) *Interior landscape plants for indoor air pollution abatement.* National Aeronautics and Space and Administration.

Wright, K., et. al. (2013) Entrainment of the human circadian clock to the natural light-dark cycle. *Current Biology.*

Wu, Y., Zhang, G., and Li, Z. (2005) MMSE results of 267 elderly persons aged above 80 in Bama. *Chinese Journal of Behavioral Medical Science.*

Xiao, Z., Xu, Q., and Yuan, Y. (1996) Solving the mystery of the status and longevity of centenarians in Bama. *Chinese Journal of Population Science.*

Xu, X., et. al. (2015) A cost-effectiveness analysis of the first federally funded antismoking campaign. *American Journal of Preventive Medicine.*

Yamashita, T. and Kunkel, S. (2010) The association between heart disease mortality and geographic access to hospitals: County level comparisons in Ohio, USA. *Social Science and Medicine.*

Yates, L., et. al. (2008) Exceptional longevity in men: modifiable factors associated with survival and function to age 90 years. *Archives of Internal Medicine.*

Zanobetti, A., et. al. (2010) Associations of PM10 with sleep and sleep-disordered breaking in adults from seven US urban areas. *American Journal of Respiratory and Critical Care Medicine.*

Zaregarizi, M., et. al. (2007) Acute changes in cardiovascular function during the onset period of daytime sleep: comparison to lying awake and standing. *Journal of Applied Physiology.*

Zhang, J., et. al. (2014) Extended wakefulness: Compromised metabolics in and degeneration of locus ceruleus neurons. *The Journal of Neuroscience.*

Zheng H., et. al. (2015) Metabolomics investigation to shed light on cheese as a possible piece in the French paradox puzzle. *Journal of Agricultural and Food Chemistry.*

INDEX

Page numbers followed by n indicate notes.

ABOUT THE AUTHORS

DR. JOHN D. DAY is a cardiologist and medical director of heart rhythm specialists at Intermountain Medical Center in Salt Lake City, Utah. He received a medical degree from Johns Hopkins and completed his cardiology training at Stanford University. Dr. Day has published more than one hundred medical studies and gives lectures throughout the world on various heart topics. He is the former president of the Heart Rhythm Society and currently serves as the Utah governor of the American College of Cardiology.

JANE ANN DAY, MA, received her master's degrees from the Georgetown School of Foreign Service and UC Santa Cruz. Her work has taken her throughout Asia, the Middle East, and North Africa, where she has empowered communities in need with self-reliance and entrepreneurial tools. Jane has published her work in international academic journals. She currently enjoys facilitating executive strategy sessions and seminars. Above all, Jane is the joyful mother of four adventurous young souls who provide her with daily opportunities to practice and teach the seven principles.

MATTHEW LAPLANTE teaches news writing and crisis reporting at Utah State University. He is the cowriter of *Inheritance: How Our Genes Change Our Lives, and Our Lives Change Our Genes* with Dr. Sharon Moalem. A former national security reporter for the *Salt Lake Tribune*, his work also has appeared in the *Washington Post* and the *Los Angeles Daily News*, and on CNN.com.